Dynamic Research Support for Academic Libraries

Dynamic Research Support for Academic Libraries

Editor
Dominika Dechiel

Dynamic Research Support for Academic Libraries
Edited by **Dominika Dechiel**

ISBN: 978-1-68117-256-9
Library of Congress Control Number: 2016934799

© 2017 by
SCITUS Academics LLC,
www.scitusacademics.com
Box No. 4766, 616 Corporate Way,
Suite 2, Valley Cottage,
NY 10989

This book contains information obtained from highly regarded resources. Copyright for individual articles remains with the authors as indicated. All chapters are distributed under the terms of the Creative Commons Attribution License, which permits unrestricted use, distribution, and reproduction in any medium, provided the original author and source are credited.

Notice

Reasonable efforts have been made to publish reliable data and views articulated in the chapters are those of the individual contributors, and not necessarily those of the editors or publishers. Editors or publishers are not responsible for the accuracy of the information in the published chapters or consequences of their use. The publisher believes no responsibility for any damage or grievance to the persons or property arising out of the use of any materials, instructions, methods or thoughts in the book. The editors and the publisher have attempted to trace the copyright holders of all material reproduced in this publication and apologize to copyright holders if permission has not been obtained. If any copyright holder has not been acknowledged, please write to us so we may rectify.

Preface

An academic library is a library that is attached to a higher education institution which serves two complementary purposes to support the school's curriculum, and to support the research of the university faculty and students. Academic librarianship is for those who are constantly intellectually curious and who can apply that curiosity to efforts that help increase the knowledge base of the institution for research, teaching, and learning. Academic libraries must determine a focus for collection development since comprehensive collections are not feasible. Librarians do this by identifying the needs of the faculty and student body, as well as the mission and academic programs of the college or university. There is a great deal of variation among academic libraries based on their size, resources, collections and services. The library is not just a repository, or a service like any other, or a place for study: it is all these things. It can also be a partner in research and in teaching, and institutions which fail to capitalise fully on this asset will find it harder to compete in the future. Higher education and academic libraries are in the age of rapid evolution. Technology, educational shifts, and programmatic changes in education mean that libraries must continually evaluate and adjust their services to meet new needs. Research and learning throughinstitutes is becoming more team-based, crossing disciplines and dependent on increasingly sophisticated and varied data. To provide valuable services in this shifting, diverse environment, libraries must think about new ways to support research on their campuses, including collaborating across library and departmental boundaries. This book, Dynamic Research Support for Academic Libraries, will enable academic libraries to provide research support for academics and graduate students and help undergraduates accomplish learning in more hands-on, in-depth ways.

Table of Contents

Chapter 1　Research data management services in academic research libraries and perceptions of librarians　　1

Chapter 2　Stereotypes Regarding Libraries and Librarians: An Approach of Romanian School and Academic Libraries　　29

Chapter 3　Strategic Management Model for Academic Libraries　　45

Chapter 4　Factors That Increase The Probability Of A Successful Academic Library Job Search　　59

Chapter 5　Current practices in library/informatics instruction in academic libraries serving medical schools in the western United States: a three-phase action research study　81

Chapter 6　Clinical and academic use of electronic and print books: the Health Sciences Library System e-book study at the University of Pittsburgh　　113

Chapter 7　Knowledge and skills for the digital era academic library　　145

Chapter 8　Library Value in the Classroom: Assessing Student Learning Outcomes from Instruction and Collections　　173

Chapter 9	University Libraries and the Development of Lecturers' and Students' Information Competencies 203
Chapter 10	Academic Education in Library and Information Management in Bulgaria 227
Chapter 11	Academic Library Services Support For Research Information Seeking 255
Chapter 12	Common Knowledge: Learning Spaces in Academic Libraries 283
Chapter 13	A university library management model for students' learning support 313
	Index 343

CHAPTER 1

Research data management services in academic research libraries and perceptions of librarians

Carol Tenopir[1], Robert J. Sandusky[2], Suzie Allard[1], Ben Birch[1]

[1]*School of Information Sciences, University of Tennessee, 451 Communications Bldg., 1345 Circle Park Drive, Knoxville, TN 37996-0341, USA*
[2]*Richard J. Daley Library, MC-234, 801 S. Morgan St., Chicago, IL 60607, USA*

ABSTRACT

The emergence of data intensive science and the establishment of data management mandates have motivated academic libraries to develop research data services (RDS) for their faculty and students. Here the results of two studies are reported: librarians' RDS practices in U.S. and Canadian academic research libraries, and the RDS-related library policies in those or similar libraries. Results show that RDS are currently not frequently employed in libraries, but many services are in the planning stages. Technical RDS are less common than informational RDS, RDS are performed more often for faculty than for students, and more library directors believe they offer opportunities for staff to develop RDS-related skills than the percentage of librarians who perceive such opportunities to be available. Librarians need opportunities to learn more about these services either on campus or through attendance at workshops and professional conferences.

1. INTRODUCTION

Science has entered a "fourth paradigm" that is more collaborative, more computational, and more data intensive (Hey, Tansley, & Tolle, 2009a) than the previous experimental, theoretical, and computational paradigms. This emerging scientific paradigm is often referred to as e-science or e-research (Hey, Tansley, & Tolle, 2009b). Increased reliance on technology in all parts of scientific endeavor, or cyberinfrastructure, and the establishment of data management and data sharing mandates by many research funding bodies[1] have motivated academic libraries to take action with regard to the shifting needs of their faculty and students and consider how best to engage in e-science through the development of library-based research data services (RDS). In the U.S. and Canada, individual large academic research libraries often lead these activities (Association of Research Libraries, 2010).

The results of investigations into (a) librarians' RDS practices in U.S. and Canadian academic research libraries and (b) the RDS-related library policies in the same type of libraries are reported here. These studies establish a baseline assessment of the RDS involvement of individual librarians as well as libraries as institutions. The results inform and enable practitioners, administrators, and educators to make strategic RDS plans in academic research libraries and guide the evolution of curricula in LIS education.

2. PROBLEM STATEMENT

The emerging need for research data management is prompting library directors to plan for additional RDS to be offered by their libraries, and at the same time many librarians are looking for opportunities to develop their RDS-related skills. But are library directors and librarians on the same page regarding RDS? In other words, do library policies in this regard align with librarians' perceptions? Misalignment can hinder effective start-up of RDS. This study focuses on the alignment issue by comparing data from library directors on RDS currently offered or planned, with data from librarians on RDS currently performed. Similarly, comparisons are made between library directors' perceptions on how their libraries are providing RDS development opportunities for staff and perceptions of librarians on the availability of such opportunities in their library. Insight on the part of the library community gained from this study could raise awareness of such misalignment, followed by corrective action leading to more efficient development of RDS.

This paper combines the findings from two surveys to answer several questions regarding North American academic research libraries and their involvement in RDS, including:

- How many academic libraries are actively engaged in RDS?

- How many academic libraries are planning to be involved in RDS in the near future?

- Are libraries offering opportunities for their librarians to gain RDS related skills?

- What specific types of RDS are being offered?

- Which are more common — *informational, consulting-type services* (for example, helping faculty and students find a place to deposit their research data or pointing to data management plan examples) or *technical, hands-on services* (for example, running an institution-housed data repository or helping researchers write data management plans)?

- Are libraries offering or thinking of offering a mix of services or concentrating on a single service?

Libraries may already be offering or planning to offer RDS and they may have developed plans to do so. However, it is the librarians who are on the front lines in terms of implementing these plans. Therefore, two separate studies were conducted; the first – the library study – surveyed directors of academic libraries in the U.S. and Canada and sought answers for the questions above. The second – the librarian study – surveyed a sample of librarians who work in academic research libraries. Librarians were surveyed in an attempt to answer such research questions as:

- How many academic librarians are currently involved with RDS in their libraries?

- Do librarians think they have the disciplinary background and training to offer RDS?

- Do librarians feel they have opportunities to learn what they need to know about RDS?

Finally, by studying both the directors of academic libraries to get official library policy and the front-line librarians who work at academic research libraries to get their perceptions and personal perspectives, the alignment between library practices and librarians' perceptions of the development opportunities open to them regarding research data management services can be evaluated. Although results from each of these individual studies have been previously reported, this article compares the results for the first time in order to determine if the practices and opportunities for training in research data service provision in academic research libraries are in alignment with the perceptions of the librarians as to their preparation and opportunities.

3. LITERATURE REVIEW

The literature related to this paper includes studies of *librarians*, *libraries*, and RDS; and papers that present case studies or recommendations for how librarians and libraries can develop RDS. In addition to surveys of the current status of RDS in libraries in several countries, current literature covers a range of recommended specific services. The literature shows that RDS or e-science services in libraries are discussed in the library literature, and are being offered by some research libraries, but are not yet being offered by most. Peters and Dryden (2011) found that the most important data services needed by researchers are mainly directional ones: grant proposal support including data management planning, locating data-related services, publication support, and specific data management assistance. Another study (Bach et al., 2012), however, found that, of the surveyed biodiversity data repositories, most only deliver low-level support for users.

Many librarians and researchers engaged with e-research have discussed the possible roles for both libraries and librarians in providing RDS (Association of Research Libraries, 2006, Council on Library and Information Resources, 2008, Gabridge, 2009,Gold, 2007, Hey and Hey, 2006 and Jones, 2009). One third of participants in a UK survey (Brown & Swan, 2007) believed that within five years "manager of datasets from e-science/grid projects" (p. 47) will be a major obligation of a librarian, with another third assigning it a secondary responsibility. MacColl (2010) advises libraries take on a more comprehensive and strategic role: libraries should be involved throughout the research process and need to be actively engaged in curating, advising, and preserving research outputs. Some additional suggested roles for libraries are to develop researchers' data-awareness, to adopt a data archiving and preservation role, and to train data librarians.

Libraries and librarians have a long way to go before realizing these roles. Potter, Cook, and Kyrillidou (2011) found that only 9 of 86 (or 10%) narrative profiles created by ARL members in the U.S. and Canada included references to "e-science/data curation and management" (p.7) as an important service supporting faculty success and scholarly communications. As part of a 2007 investigation into all of the types of support provided to researchers by 134 U.S. and Canadian academic health science libraries, Cheek and Bradigan (2010) found that just 12.2% of these libraries provided support for "data curation". About half of the respondents to ARL's 2009 North American e-science survey (Association of Research Libraries, 2010) had on-campus support units for scientific research data; however, the Data Working Group at Cornell University Library discovered that few university libraries were actually involved in research data curation (Steinhart et al., 2008).

A major aspect of data curation is preservation work. A 2009 survey of European data managers, of which nearly three-quarters (73%) were employed in libraries, found that the top three reasons for research data preservation included accountability for publicly funded research, inspiration for scientific advancements, and reanalysis of previously generated data (Kuipers & Van der Hoeven, 2009, p. 37). A majority of data managers, including those not employed by libraries, reported that their institutions have a policy for preservation of research data. These data managers did not report on the percentage of the data that was shared with other researchers. One Australian study found that data sharing is not a priority for many researchers (Markauskaite, Kennan, Richardson, Aditomo, & Hellmers, 2012). In this study, 864 researchers at seven Australian Universities were surveyed; 50% of participants did not allow access to any of their data and only 9% provided access to all of their data.

3. Literature review

Research institutions have a responsibility to offer researchers educational and support services relating to data management and to encourage data sharing, in addition to providing policies and structure for research data preservation (Tenopir, Birch, & Allard, 2012). There is a need for more tailored and streamlined data services, but the identification of researchers' needs is difficult due to the complex nature of research data.Carlson (2012) at Purdue University Libraries, found that among and within fields of study there are disparities in the way that data curation is conceptualized and communicated. These variations make it challenging for librarians to understand the needs of researchers.

There are also many other challenges that librarians and libraries face in RDS development. Corrall, Keenan, and Afzal (2013) found clear evidence that development of specialized RDS is often constrained by knowledge and skills gaps among library staff and a lack of confidence in their expected roles in RDS. In a small-scale survey of New Zealand academic and college library managers, Brown (2010) found that there was little direct involvement in providing RDS but that libraries were participating in local steering groups, performing institutional planning, and involved in policy development both within and between academic institutions. This survey found that funding, librarian training, marketing, and uncertain demand from researchers and students were barriers to successfully providing RDS (Brown, 2010). Similarly, Creamer, Morales, Crespo, Kafel, and Martin (2012) found that health and science librarians have a high level of interest in developing a range of RDS skills, but often lack the skills needed to effectively provide RDS. More than half of the libraries were creating a "library strategic plan or policy for data management" (p.21), although they faced "serious barriers" to engaging in e-science, including funding for personnel and equipment and lack of broader institutional support, as well as "territorial struggles" between various other departments within the institution (p.23).

Recently, there have been tools developed and recommendations made to help overcome the challenges faced by librarians and libraries. The result of Carlson's (2012) terminology variation study was the development of the DCP Toolkit (http://datacurationprofiles.org). This tool enhances the data reference interview and enables librarians to connect with and discover the data needs of researchers. The detailed profiles in the DCP Toolkit provide insight into the data management language utilized by researchers in different fields. A number of studies have also acknowledged the importance of educating library staff about data curation and management services. Many library staff members have collection experience related to traditional materials, but may require training in relation to selecting and compiling data for inclusion in repositories. Research libraries have campus-wide faculty relationships and are proficient at developing conventional collections, giving them a competitive advantage in establishing a university's scientific data collection; however, Newton, Miller, and Bracke (2010) found that additional training is needed to build up an institutional data repository. Libraries need to utilize their professional connections with campus faculty, as well as faculty and staff at other institutions to collaborate and develop more skills in identifying appropriate materials.

4. METHODOLOGY

In order to compare like-to-like (that is, librarians in research libraries with policies of research libraries), the library data reported here are a subset of a larger study that surveyed all types and sizes of academic libraries in the U.S. and Canada. The full study sought to answer, in addition to the research questions addressed here, what sizes and types of academic libraries are most involved in RDS and how involvement varies by type and size of academic library. The full results were reported in Tenopir et al., 2012.

4. Methodology

In the full study, survey responses were received from 223 library directors. In order to investigate the effects of variances in sizes and types of libraries, four demographic characteristics of the parent institution were used: number of full-time equivalent (FTE) students (less than 5000 vs. 5000 or more), number of tenure-track and tenured faculty (less than 100 vs. 100 or more), number of National Science Foundation (NSF) grants typically awarded per year (none vs. some), and type of institution (research or doctoral vs. baccalaureate vs. associate's).

Not surprisingly, academic research libraries, larger schools, and those receiving more NSF grants were more likely to be offering or planning to offer research data management services than other types of academic libraries (Tenopir et al., 2012).

Comparing results by these demographics also uncovered differences in methods libraries were using to develop staff capacity for RDS. Libraries at institutions with high enrollments, those with a large faculty, and those at research institutions were all more likely to have already reassigned or to be planning to reassign existing staff than libraries at other institutions.

Finally, there were considerable differences in library engagement with RDS. Libraries at institutions with high enrollment, larger faculty size, and at research institutions were more likely than libraries at smaller schools to be involved in things such as managing RDS technology infrastructure, planning RDS skills development opportunities for staff, and collaborating with other units on campus (Tenopir et al., 2012).

Because of these differences, the authors decided to examine in more depth only the results from libraries in research or PhD-granting institutions and to compare the results with another survey of individual academic librarians who work in research universities.

As in the academic library study, the librarian data reported here also represent a subset of a larger study. In the larger study, librarians employed by ARL member libraries were surveyed if their area of responsibility seemed likely to currently, or in the future, include RDS. The full study sought to answer how librarians' opinions of their preparedness to provide RDS, their library's support for their professional RDS development, the importance of RDS for libraries and their associated institutions, and the contributing or inhibiting factors for librarian involvement in RDS, varied with their current degree of engagement with RDS. The results of this part of the full study were reported in Tenopir, Sandusky, Allard, and Birch (2013).

The full results indicated consensus that the absence of RDS would adversely affect the institution's perception of the library in terms of relevance and prestige, that provision of RDS would augment the institution's research impact, and that the absence of RDS would put the institution at a disadvantage for grants. In addition, participants strongly rejected the idea that RDS would be a distraction and the idea that RDS are unnecessary and strongly affirmed that RDS fits the traditional role of librarians as stewards of scholarship (Tenopir et al., 2013).

The current analysis compares, for the first time, the frequency of RDS provision in academic research libraries with the services offered by the librarians.

The full library study was distributed to a stratified random sample of 351 library directors who are members of a panel organized by the Association of College and Research Libraries (ACRL). Each of these directors had agreed to participate in several ACRL surveys on assorted topics over the course of a year. A total of 221 of these ACRL directors responded, for a response rate of 63%. Surveys were initially distributed in November 2011, with a follow-up in January 2012 (Table 1).

4. Methodology

Table 1. Distribution to ACRL panel members.

Classification	Panel members	Responses	Response rate
Associate-degree granting	116	68	59%
Baccalaureate-degree granting	93	54	58%
Doctorate-granting	142	99	70%
Totals	351	221	63%

A separate distribution to several libraries in the University of California (UC) system yielded two additional responses. The number of invitations sent in the UC system, and therefore the exact response rate is unknown, although there are ten campuses in the UC system. The final dataset contains responses from 223 academic libraries. This paper focuses on the 101 responses from research/doctorate granting institutions (the 99 from ACRL distribution and the two UC universities; see Tenopir et al. (2012) for a full analysis of all 223 library responses).

In order to compare official library policy with the perspective of the librarians who work in academic research libraries, a separate survey was sent to a sample of academic librarians in the U.S. and Canada. The survey was sent to librarians who work in the 115 academic libraries that belong to the ARL. Between April 2011 and August 2011, a total of 948 invitations were sent to a sample of ARL librarians who work as subject librarians, metadata librarians, e-science librarians, or data librarians. This survey had 222 responses: a response rate of 23%.

In November 2011 and February 2012 a separate invitation was sent to librarians working in two libraries in the UC system, and to librarians whose ACRL library director volunteered to distribute the survey to their staff. This yielded 80 additional responses. The exact number of invitations sent in this method, and therefore the response rate, is unknown. The final dataset includes responses from both the initial distribution and this second distribution for a total of 302 librarians.

All of the respondents to the librarian survey work in comprehensive research-extensive institutions, while libraries in the libraries survey included associate, baccalaureate, and doctorate degree-granting institutions. Therefore, in order to remove a confounding factor from comparisons between the two survey results, only libraries at doctorate degree-granting institutions are included in this analysis. That way, although it is not known whether librarian respondents come from the same institutions that responded to the libraries survey, the official policies of academic research libraries can be better compared with the perceptions of librarians who work in that type of institution.

Questions to both library directors and librarians covered specific RDS offered or planned to be offered in their institutions, as well as opportunities for professional development on RDS issues for the professional staff. Half of these questions concerned informational or consulting RDS and half were about a greater level of involvement with technical/hands-on RDS. Informational/consulting services cover a wide range of services, from consulting on data management plans through discussing RDS with others:

- Consulting with faculty, staff, or students on data management plans.

- Consulting with faculty, staff, or students on data and metadata standards.

- Outreach and collaboration with other RDS providers either on or off campus.

- Providing reference support for finding and citing data or datasets.

- Creating Web guides and finding aids for data, datasets, or data repositories.

- Discussing RDS with other librarians, or other people on campus, or RDS professionals.

4. Methodology

The technical or hands-on services show another level of involvement with RDS:

- Providing technical support for RDS systems (e.g., a repository, access, and discovery systems).
- Deaccessioning or deselection of data or datasets for removal from a repository.
- Preparing data or datasets for deposit into a repository.
- Creating or transforming metadata for data or datasets.
- Identifying data or datasets that could be candidates for repositories on or off campus.
- Directly participating with researchers on a project (as a team member).

Library directors were asked whether each of the RDS were currently offered or planned to be offered in the future through the library. The answer choices were:

1. Not available, and we currently have no plans to offer it.
2. Not available, but we plan to offer it in more than 24 months.
3. Not available, but we plan to offer it within 13–24 months.
4. Not available, but we plan to offer it within 12 months.
5. Our library currently offers this service.

Librarians were asked how frequently they performed each of the RDS. Their answer choices were:

1. Never performed.
2. Performed a few times a year.
3. Performed about once a month.

4   Performed about once a week.

5   Performed daily.

In cases where the respondent indicated that service was provided to both faculty and students, separate details were given for faculty and students.

5. RESULTS

Providing reference support for finding and citing data or datasets is the most common of currently-offered or planned-to-be-offered informational RDS in academic research libraries, with nearly half currently offering this service and another third planning to within the next two years (Table 2). That means that almost 83% of these libraries will offer this service within the next two years. No other data informational service is currently offered by a majority of libraries, but consulting on data management plans, consulting on metadata creation, creating guides, and discussing RDS with patrons are planned in a majority.

Outreach and collaboration with other RDS providers either on or off campus is the least commonly offered information service now and least likely to be in the planning stages, although this may merely mean that the library doesn't need to collaborate in order to offer RDS.

The following two tables (Table 3A and Table 3B) show how often informational RDS are performed by librarians in academic research libraries for faculty (Table 3A) or for students (Table 3B). On average, informational RDS are currently performed by these librarians never or only a few times a year, with helping faculty find relevant data or datasets the most frequently offered service. Helping students find data or datasets is the most frequently offered RDS for students, followed closely by creating library guides to data services. Table 3A and Table 3B also show that in each case in which a service is performed for faculty or students, it is performed more often for faculty than for students.

5. Results

Table 2. Informational or consulting RDS currently offered by the library or planned to be offered in the future [library study].

Table 3A. Informational or consulting RDS currently performed by librarians for faculty [librarian study].

RDS	Never	Few times/ year	Once/ month	Once/ week	Daily
Create guides (n = 262)	44.0%	38.0%	10.0%	6.0%	2.0%
Find data (n = 255)	32.0%	41.0%	16.0%	6.0%	5.0%
Consult meta (n = 261)	52.0%	35.0%	7.0%	5.0%	1.0%
Consult DMP (n = 265)	55.0%	29.0%	12.0%	3.0%	0.0%

Table 3B. Informational/consulting RDS currently performed by librarians for students [librarian study].

RDS	Never	Few times/ year	Once/ month	Once/ week	Daily
Create guides (n = 239)	53.0%	33.0%	8.0%	4.0%	2.0%
Find data (n = 233)	36.0%	39.0%	11.0%	9.0%	5.0%
Consult meta (n = 240)	74.0%	20.0%	3.0%	2.0%	1.0%
Consult DMP (n = 239)	81.0%	13.0%	5.0%	0.0%	1.0%

Although over half of all research libraries in the libraries survey do not officially consult with others on campus or beyond the campus for RDS, individual librarians are slightly more likely to report that they collaborate on RDS (Table 4). Almost half of librarians engage in outreach, that is, collaborating with other RDS providers on campus. Working beyond their campus is more common, and 60% of the librarians report they participate in working groups or other professional groups about RDS. This participation may lead to a growth in RDS in libraries in the future.

In general, technical RDS are currently less frequently offered in libraries than are informational or consulting RDS, but only slightly so (Table 5). Roughly half of these academic research libraries offer or plan to offer most of the other technical services within two years, including (in order of frequency) within two years: providing technical support for a data repository, identifying datasets to incorporate into an institutional repository, providing librarians to serve as team members on e-science projects, creating metadata for datasets, and preparing data for deposit. Only deaccessioning datasets is not offered or planned by a majority of academic research libraries.

Table 4. Informational or consulting RDS currently performed by librarians with others [librarian study].

RDS	Never	Few times/year	Once/month	Once/week	Daily
Outreach (n = 218)	53.0%	34.0%	8.0%	5.0%	1.0%
Groups (n = 218)	40.0%	35.0%	15.0%	8.0%	2.0%

Librarians also report that they perform technical RDS less frequently than informational or consulting RDS, although the questions were slightly different in the librarian survey. Neither identifying datasets for inclusion in a repository nor serving as an e-science team member for faculty (Table 6A) or for students (Table 6B) is offered by a majority of these librarians. For those who do offer the services, a "few times per year" is the mostly likely frequency.

Table 6C shows the technical services offered by librarians regardless of for whom. Very few librarians offer these services, although they may be covered by a single data services librarian in those institutions that support the service. Creating metadata for data is the most commonly offered service, but most frequently only a few times per year. There is wide variation: 207 of 226 respondents reported never performing deaccession or deselection of data or datasets from repository, although one respondent reported performing this service daily.

Table 5. Technical RDS currently offered by the library or planned to be offered in the future [library study].

RDS	No plans	>24 months	13–24 months	<12 months	Has service
Team member (n = 100)	49.0%	11.0%	5.0%	8.0%	27.0%
Identify data (n = 98)	44.9%	10.2%	12.2%	15.3%	17.3%
Create meta (n = 97)	50.5%	10.3%	16.5%	6.2%	16.5%
Prepare data (n = 100)	52.0%	9.0%	13.0%	11.0%	15.0%
Deaccession (n = 100)	73.0%	8.0%	10.0%	6.0%	3.0%
Tech support (n = 100)	43.0%	12.0%	16.0%	11.0%	18.0%

Table 6A. Technical RDS currently performed by the librarian with faculty or staff [librarian study].

RDS	Never	Few times/ year	Once/ month	Once/ week	Daily
Identify data (n = 256)	52.0%	31.0%	12.0%	4.0%	1.0%
Team member (n = 256)	65.0%	26.0%	5.0%	3.0%	1.0%

Table 6B. Technical RDS currently performed by the librarian with students [librarian study].

RDS	Never	Few times/ year	Once/ month	Once/ week	Daily
Identify data (n = 233)	71.0%	21.0%	4.0%	3.0%	1.0%
Team member (n = 231)	82.0%	13.0%	3.0%	1.0%	1.0%

Table 6C. Technical RDS currently performed by the librarian on data or datasets [librarian study].

RDS	Never	Few times/ year	Once/ month	Once/ week	Daily
Tech support (n = 224)	76.0%	15.0%	4.0%	2.0%	4.0%
Deaccession (n = 226)	92.0%	7.0%	1.0%	0.0%	0.0%
Prepare data (n = 228)	71.0%	21.0%	6.0%	1.0%	2.0%
Create meta (n = 229)	67.0%	20.0%	6.0%	3.0%	4.0%

The library study also explored whether libraries provided opportunities for staff to develop skills related to RDS. Just under one-third (31 of 99) replied that they provide some opportunities. Those that replied yes also reported which of the following opportunities were provided:

1 In-house staff workshops or presentations.

2 Taking courses related to RDS.

3 Attending conferences or workshops elsewhere related to RDS.

Library directors could select all that applied to their organizations. Providing conference opportunities was the clear favorite, with 94% of libraries offering attendance at conferences to their librarians (Table 7). Slightly more than half of libraries reported supporting attendance at courses related to improving RDS skills.

Table 7. Percentage of libraries providing specific opportunities for staff to develop RDS skills. Library directors were allowed to select all that applied [library study].

Opportunities for RDS skills (n = 99)	Percentage
Training	45%
Courses	58%
Conferences	94%

Among libraries that currently offer any training opportunities, Table 7 shows the percentages of libraries offering each opportunity. However, when all libraries are considered together, including those that do not offer services, the percentages are much lower: 29% for conferences, 18% for courses, and 14% for in-house training.

Some librarians feel they have opportunities for learning. Just under half (47%) of the 219 librarians who answered this question felt they had the opportunity for at least one type of RDS skills development (Table 8). The most common opportunity was support for attendance at conferences (65%), followed by courses elsewhere (53%), and, less often, training at their library (32%).

Table 8. Percentage of librarians agreeing that their library provides specific opportunities to develop RDS skills [librarian study].

Agreement with opportunities for RDS skills	Percentage
Training (n = 218)	32%
Courses (n = 216)	53%
Conferences (n = 217)	65%

6. DISCUSSION

The most commonly offered or planned informational RDS, finding and citing datasets, (Table 2) is a service that simply extends a familiar library reference service into the realm of data. At the other extreme, the least commonly offered information service is outreach and collaboration with other RDS providers. With the extensive hardware, software, and educational components needed for effective RDS, it is somewhat disheartening that so few research libraries are collaborating with others.

According to librarians who work in academic research libraries, RDS are being performed never or only a few times a year (Table 3A and Table 3B). Keeping in mind that Table 2 takes into account planned availability as well as current availability, this result is not surprising. Growth in current performance of RDS by librarians can be expected to follow growth in current availability of RDS by libraries.

Considering technical RDS, this type of service is less available in libraries than are informational RDS (Table 4). The picture is also likely to change significantly in the near future, as only one service (deaccessioning or deselection of data or datasets for removal from a repository) is neither in the plans nor currently being offered by most libraries. Perhaps the preservation aspect of repositories is considered inconsistent with deselecting data that is put into a repository.

There appears to be somewhat of a mismatch between what academic research library directors believe they offer to their librarians and what the librarians themselves perceive to be available to them in the way of RDS training opportunities (Table 8). Nonetheless, these results portend well for the future of RDS, as there are clearly some opportunities for training of librarians in RDS skills.

Library directors and librarians who are aware of research data management issues or are currently involved in RDS are more likely to respond to the survey, so results may show an inflated picture of research library involvement in RDS. Also, this is a dynamic topic and plans may change at some of the responding institutions. Responses to the libraries survey concern library policy, with the unit of analysis at the library institutional level. Responses to the librarian survey concern individual perceptions and opinions, with the unit of analysis at the individual librarian level. It is not possible to know how many of the librarian-respondents work at the institutions that responded to the libraries survey. It is also not possible to know how many individuals from the same institution responded. In addition, since this paper focused on libraries and librarians at research or doctorate-granting institutions, the results are not generalizable to academic libraries and librarians at other types of schools such as baccalaureate and associate degree-granting institutions. And, finally, as with any survey, responses are self-reported and are assumed to be accurate and truthful, although this cannot be verified.

7. CONCLUSION

It is clear that some academic research libraries are offering a variety of research data management services and more plan to do so within the next two years. Most commonly these services are extensions of traditional informational or consultative services, such as helping faculty and students locate datasets or repositories. A small, but growing, number of libraries are becoming more involved with research data, from helping with data management plans to preparing and preserving research data for deposit in data repositories.

Many of the librarians who work in academic research libraries feel they have the subject knowledge necessary to help their constituents with research data services, but need the opportunity to take advantage of continuing education. Whether consultative or hands-on services, librarians need opportunities to learn more about these services either on their own campus or through attendance at workshops and professional conferences.

Working with others on campus, as both teachers and joint learners of research data service specifics, will help the library play a shared role in building the future of research data at their universities.

The comparisons drawn here between library policy on RDS and the perceptions of front-line librarians as they implement this policy, indicates some misalignment. However, that is to be expected, as most libraries are in the early stages of making RDS available. Increased awareness of this issue within the academic library community is likely to result in more effective development of RDS.

ACKNOWLEDGMENTS

This work was supported by Data Observation Network for Earth (DataONE), National Science Foundation award #0830955 under a Cooperative Agreement. We would like to thank Graduate Research Assistant Madison Langseth for her careful work on helping us respond to the peer reviewers' suggested revisions.

REFERENCES

1. Association of Research Libraries (2006). To stand the test of time: Long-term stewardship of digital data sets in science and engineering. Washington, D.C.: Association of Research Libraries. Retrieved from http://arl.nonprofitsoapbox.com/storage/documents/publications/digital-data-report-2006.pdf

2. Association of Research Libraries (2010). E-science and data support services: A study of ARL member institutions.Washington, DC: Association of Research Libraries. Retrieved from http://www.arl.org/storage/documents/publications/escience-report-2010.pdf

3. Bach, K., Schafer, D., Enke, N., Seeger, B., Gemeinholzer, B., & Bendix, J. (2012). A comparative evaluation of technical solutions for long-term data repositories in integrative biodiversity research. Ecological Informatics, 11, 16–24.

4. Brown, E. (2010). I know what you researched last summer: How academic librarians are supporting researchers in the management of data curation. The New Zealand Library & Information Management Journal, 52(1), 55–69. Retrieved from http://www. lianza.org.nz/sites/lianza.org.nz/files/nzlimj_vol_52_issue_no_1_oct_2010.pdf

5. Brown, S., & Swan, A. (2007). Researchers' use of academic libraries and their services: A report commissioned by the Research Information Network and the Consortium of Research Libraries. London, UK: Research Information Network and Consortium of Research Libraries in the British Isles. Retrieved from http://www.rin.ac.uk/our-work/ using-and-accessing-information-resources/researchers-use-academic-librariesand-their-serv

6. Carlson's, J. (2012). Demystifying the data interview: Developing a foundation for reference librarians to talk with researchers about their data. Reference Services Review, 40(1), 7–23

References

7. Cheek, F. M., & Bradigan, P. S. (2010). Academic health sciences library research support. Journal of the Medical Library Association, 98(2), 167–171.

8. Corrall, S., Keenan, M. A., & Afzal, W. (2013). Bibliometrics and research data management services: Emerging trends in library support for research. Library Trends, 61(3), 636–674

9. Council on Library and Information Resources (2008).No brief candle: Reconceiving research libraries for the 21st century. Washington, DC: Council on Library and Information Resources. Retrieved from http://www.clir.org/pubs/reports/pub142/pub142.pdf

10. Creamer, A., Morales, M. E., Crespo, J., Kafel, D., & Martin, E. R. (2012). An assessment of needed competencies to promote the data curation and management librarianship of health sciences and science and technology librarians in New England. Journal of eScience Librarianship, 1(1), 18–26

11. Gabridge, T. (2009). The last mile: The liaison role in curating science and engineering research data. Research Library Issues: A Bimonthly Report from ARL, CNI, and SPARC, 265. (pp. 15–21). Retrieved from http://publications.arl.org/rli265/16

12. Gold, A. (2007). Cyberinfrastructure, data, and libraries, part 2: Libraries and the data challenge: Roles and actions for libraries. D-Lib Magazine, 13(9/10). Retrieved from http://www. dlib. org/dlib/september07/gold/09gold-pt2.html

13. Government of Canada (2013). Tri-agency open access policy. Retrieved from http://www. nserc-crsng.gc.ca/NSERC-CRSNG/policies-politiques/OpenAccessFAQLibreAcces FAQ_eng.asp#8

14. Hey, T., & Hey, J. (2006). E-science and its implications for the library community. Library Hi Tech, 24(4), 515–528

15. Hey, T., Tansley, S., & Tolle, K. (2009a). The fourth paradigm: Data-intensive scientific discovery. Redmond, WA: Microsoft Corporation. Retrieved from http://research.microsoft.com/ en-us/collaboration/fourthparadigm/4th_paradigm_book_complete_lr.pdf

16. Hey, T., Tansley, S., & Tolle, K. (2009b). Jim Gray on eScience: A transformed scientific method. In T. Hey, S. Tansley, & K. Tolle (Eds.), The fourth paradigm: Data-intensive scientific discovery (pp. xix–xxxiii). Redmond, WA: Microsoft Corporation. Retrieved from http://research.microsoft.com/en-us/collaboration/fourthparadigm/4th_ paradigm_book_complete_lr.pdf

17. Jones, E. (2009). Reinventing science librarianship: Themes from the ARL-CNI forum. Research library issues: A bimonthly report from ARL, CNI, and SPARC, 262. (pp. 12–17). Retrieved from http://publications.arl.org/rli262/13

18. Kuipers, T., & Van der Hoeven, J. (2009). Insight into digital preservation of research output in Europe: Survey report (D3.4). Didcot, UK: PARSE.Insight. Retrieved from http:// www.parse-insight.eu/downloads/PARSE-Insight_D3-4_SurveyReport_final_hq.pdf

19. MacColl, J. (2010). Library roles in university research assessment. LIBER Quarterly, 20(2), 152–168. Retrieved from http://research-repository.st-andrews.ac.uk/bitstream/ 10023/1677/1/ MacColl2010LIBERQuarterly20LibraryRoles.pdf

20. Markauskaite, L., Kennan, M. A., Richardson, J., Aditomo, A., & Hellmers, L. (2012). Investigating eResearch: Collaboration practices and future challenges. In A. Juan, T. Daradoumis, M. Roca, S. Grasman, & J. Fauli (Eds.), Collaborative and distributed e-research: Innovations in technologies, strategies and applications (pp. 1–33).

References

21. Newton, M. P., Miller, C. C., & Bracke, M. S. (2010). Librarian roles in institutional repository data set collecting: Outcomes of a research library task force. Collection Management, 36(1), 53–67.

22. Peters, C., & Dryden, A. R. (2011). Assessing the academic library's role in campus-wide research data management: A first step at the University of Houston. Science & Technology Libraries, 30(4), 387–403

23. Potter, W. G., Cook, C., & Kyrillidou, M. (2011). ARL profiles: Research libraries 2010. Washington, D.C.: Association of Research Libraries. Retrieved from http:// www.arl.org/storage/ documents/ publications/arl-profiles-report-2010.pdf

24. Steinhart, G., Saylor, J., Albert, P., Alpi, K., Baxter, P., Brown, E., et al. (2008). Digital research data curation: Overview of issues, current activities, and opportunities for the Cornell University Library. A report of the Cornell University Library Data Working Group. Retrieved from http://ecommons.library.cornell. edu/bitstream/ 1813/10903/ 1/DaWG_WP_final.pdf

25. Tenopir, C., Birch, B., & Allard, S. (2012). Academic libraries and research data services: Current practices and plans for the future. (An ACRL white paper). Chicago, IL: Association of College and Research Libraries. Retrieved from http://www.ala.org/acrl/sites/ala.org.acrl/files/content/publications/whitepapers/Tenopir_Birch_Allard.pdf

26. Tenopir, C., Sandusky, R. J., Allard, S., & Birch, B. (2013). Academic librarians and research data services: Preparation and attitudes. IFLA Journal, 39(1), 70–78

27. University of Minnesota Libraries (2011). Funding agency and data management guidelines. Minneapolis–St. Paul, MN: University of Minnesota. Retrieved from https://www.lib. umn.edu/ data-management/funding

CHAPTER 2

Stereotypes Regarding Libraries and Librarians: An Approach of Romanian School and Academic Libraries

Maria Micle

West University. Faculty of Political Sciences, Philosophy and Communication Sciences, Timișoara, Romania

ABSTRACT

The stereotypes on librarians in Romania generally imply the following features: women, glass-wearers, suspicious, mostly grumpy, etc. These labels are still strong because many libraries, particularly school libraries, are not digitised and, therefore, supply traditional services. To this, we should add the reservations about the library space which most potential users avoid because of its communist image: a boring, cold, dusty space with old books. Positive attributes are equally used, in contradiction with negative ones: librarians are communicative, kind, good readers, supportive in finding information–all of which show that the interactions between librarians, the public and the way information is conveyed in documentary services have diversified thanks to the development of new communication and information technologies thus allowing the re-invention of library identity through information culture, management policies, spaces, access to collections and services. To analyse the measure in which traditional stereotypes about libraries still act or have changed in the mind of the Romanian public and librarians on library trade, we used as research methods the group-focus, the interview, thus attempting to extract from the data we collected representational models by overlapping common points in the respondents views. We also found stereotypes on blogs, in anecdotes, and fiction.

KEYWORDS

Stereotypes; Librarians; School Libraries; Romanian Academic Libraries

1. INTRODUCTION

Librarians are considered "cultural heroes" that are discrete, sometimes ignored, but indispensable in the life of a community when it is about education or the preservation, organisation, or reuse of information as cultural memory: this is why we attempt at reconstructing their stereotypical portraits. Stereotypes about libraries and librarians are used all over the world, so they are in Romania; they resist change despite the fact that the interaction between librarian and public, as well as the conveyance of the information within documentary services have diversified thanks to the development of new communication and information technologies: on-line reference services, e-mail, web page information – thus re-inventing the identity of librarians from the perspective of information culture, of management policies, of space, of access to collections and services. Our study aims at finding out if Romanian mentality regarding libraries has started to change or not. We find the ways librarians from all over the world have chosen to change the image of the traditional library interesting and ingenious.

In Romania, the stereotypes generally conveyed and popularised about the trade of librarian are woman, glasswearer, suspicious, mostly grumpy, etc. These labels are still strong also because many libraries, particularly school libraries, are not digitised and, therefore, supply traditional services. To this, we should add the reservations about the library space which most potential users avoid because of its communist image: a boring, cold, dusty space with old books. Prejudiced ideas about the trade are detrimental and keep the public far from libraries as sources of information and documenting and from choosing the trade of librarian as a future job and/or career.

2. METHODOLOGY

The research methods we have used to investigate this topic – identifying stereotypes related to Romanian school and academic libraries and librarians were focus-group and direct interview. We have used these qualitative methods on both library users and librarians. It would be also relevant to know what librarians think about how other people see them and if there are any professional self-stereotypes – which is the topic of another study.

We organized five focus-groups: three on library users and two on librarians. The five meetings took place in three different cities of Romania (Timișoara, Buziaș, and Reșița) and the participants were very composite from the point of view of age, environment (urban/rural), and information concerns. The results presented in this paper follow the analysis of the data from the three focus-groups whose target groups were school library users: the participants of the focus-groups 1 and 2 were MA students in Communication of the West University of Timișoara (18 and 9 people, respectively); focus-group 3 consisted of terminal high-school students from the Grammar High-school of Buzias (15 people). We completed the information thus obtained with stereotypical ideas selected from discussions and posts on blogs, from anecdotes, and from fiction.

To investigate the opinions of the library users, we used a discussion guide and we asked for free characterizations of both libraries and librarians. The guide contained the following questions:

a) What do you have in mind when you say library / librarian?

b) Why do you choose / avoid going to the library?

c) Are most librarians you know men / women?

d) Do you think female librarians are more competent than male librarians?

e) What is the average age of the librarians you know: 20-35, 35-45, 45-65? f) Name at least 3 positive and at least 3 negative attributes of a librarian.

3. STEREOTYPE: DEFINITION AND HISTORY

Stereotypes are knowledge structures that associate members of different social categories with certain attributes that make up the content of the stereotype. Stereotypes are widely spread socially, i.e. people belonging to a society understand the content of a stereotype in the same way. "Between reality and arbitrary, people often choose the category that best meets despise" (BOURHIS, LEYENS, 1997: 5). Cultural stereotypes differ from personal stereotypes which reflect personal convictions.

Romanian stereotip comes from French stéréotype < cf. Greek stereos 'solid', typos 'character' (cf. Dexonline: accessible at www.dexonline.ro). The term was first used in printing, in 1789, meaning 'a relief printing plate cast in a mould made from composed type or an original plate'; stereotyping meant 'mass reproduction of a fixed model' (SAMSON, 2011: 12). As in the case of other technical terms – graphics, photography (e.g., cliché), and printing – the term stereotype got to be overtaken and attributed other meanings attested by usage. Thus, starting from the idea of 'immobility', of 'matrix', the term stereotype got to be well known in the field of socio-psychological sciences where it was introduced by in 1922 by the American publicist and researcher Walter Lippman in his book Public Opinion, pointing to the rigidity of the human views of social groups, in particular. Lippman used the term stereotype to name "the images in the human mind that mediate the relationship between the humans and the environment", ready-made representations of pre-existing cultural schemes each individual uses to filter the environment (HERSCHEBERG-PIERROT, 1997, apud SAMSON, 2011: 13). From the point of view of social psychologist, stereotypes are defined as "cognitive structures stored in humans' memory that affect both perception and behaviour at group level" (CERNAT, 2005: 26).

4. POSITIVE AND NEGATIVE STEREOTYPES RELATED TO LIBRARIANS AND LIBRARIES

Stereotypical judgements can be positive and negative, more or less real, and vary at different times or in the context of certain events (BOURHIS, LEYENS, 1997: 5).

A librarian's profile worldwide† :

- It is rather a she, between two ages;
- Wears glasses and is dressed formally and old-fashioned;
- Is shy, not friendly (Librarian Stereotypes. Librarian Portrayals. iAccesible at: http://whatisalibrarian.blogspot.ro/2012/10/librarian-portrayals-introduction.html.);

- Is privileged because he/she has plenty of time to read;
- Does "nothing" when there are no readers

Though social psychology characterises stereotypes as fixed and rigid, stereotypes associated to librarians are easily blown off when the users come to a library and interact directly with professional librarians, characterising them as sociable, innovative, adaptable to current information requirements in a "society of knowledge" based on knowledge technologies: "Show librarians that are adaptable, cutting edge, helpful, hip, innovative, professional, and tech-savvy" (Annis Lee ADAMS, Jenny EMANUEL et al. Rebranding the Librarian Profession. HLAA Annual Conference, November 10, 2007).

Source: http://librariotypes.wordpress.com/2012/02/16/librariotypes-presents-how-people-view-my-profession-memes/

5. STEREOTYPICAL ATTRIBUTES FOR ROMANIAN LIBRARIES AND LIBRARIANS

The Romanian info-documentary system is made up of the following types of libraries: academic, school, school documenting and information centres – for the educational system, and county, municipal, and communal libraries – for the local communities, and specialised libraries. In addition, Romania's National Library is the coordinator of the legal book depot, controls national bibliography, and attributes ISBN and ISSN numbers to the different editors/publishers.

We focused mainly on the libraries of the educational system in an attempt to analyse the stereotypes used by the users about Romanian school and academic libraries. The analysis of the data we collected shows that these stereotypes correspond, in general, to the international stereotypes; however, we found a few features of Romanian libraries and realities.

Stereotypes about Romanian traditional libraries and librarians are very resistant to change and own a genuine emotional load even though the interaction between librarian and the public and the conveyance of information within documenting services have diversified through the development of new communication and information technologies – on-line reference services, e-mail, web page information – all of which helped reinventing the identity of libraries from the perspective of information culture, management policies, spaces, and access to collections and services.

6. RESULTS OF THE DISCUSSIONS WITHIN USER FOCUS-GROUPS

Multiple attributes and particularly the consequent repetition of some of them lead to a clearer shaping of the profession and of the institution of the library. Table 1 below presents only the attributes mentioned repeatedly by the participants to the discussion except for the very subjective and the very sporadic ones.

Table 1. Multiple attributes associated with libraries by focus-group participants

Focus-groups 1, 2. Participants: MA students (about academic libraries)		Focus-group 3. Participants: students, teachers (about school libraries)	
Positive	Negative	Positive	Negative
information	boredom	useful information	
knowledge	monotony	knowledge	
silence			silence
printed book	not enough copies	culture	old printed book
free access to information	few individual study rooms		
Internet available	few individual work cabinets		
organized area	suspicion		
meeting place	a place for the swots, nerds		
relaxation	dust	intimacy	lost time

We need to mention that, in Tables 1 and 2, the attributes set in the positive or negative column were ranged as considered by the groups of users of academic and school libraries: this is why there are situations apparently paradoxical in which the same attribute can be found in the positive category according to the students and in the negative one according to the high school students. For instance, *silence* specific to libraries is a positive feature for the students and a negative feature for the high school pupils which could be explained from the perspective of

6. Results of the discussions within user focus-groups

the age: students appreciate the conditions necessary for intellectual work, while teenagers prefer socialising and noise even in a library. The same for printed book: students are content to find and read printed books in a library since they are accustomed to the diversity of the informational supports, while high school pupils consider printed books obsolete in the school library and wish they had more electronic resources. In the case of the students, the fact that the staffs of libraries is mainly female is negative, while high school pupils consider it positive, maybe because of their attachment to female figures (mothers): "We are raised by women and then we tend to generalise our attachment for maternal figures, to women, in general" (CERNAT, 2005: 104).

Table 2. Multiple attributes associated with libraries by focus-group participants

Focus-groups 1, 2. Participants: MA students (about academic libraries)		Focus-group 3. Participants: students, teachers (about school libraries)	
Positive	Negative	Positive	Negative
professionals	book worms	calm	book worms
talkative	woman	woman	uninterested
assist in finding information	aged	sociable	unavailable (sometimes)
nice	irritated		grumpy
read a lot	ironical	reading lover	
are passionate about culture	mystical, enigmatic		bored
	glass-bearers		

Knowing the real attributes of the professional group of librarians help removing preconceived ideas: this is why direct interaction and observation of groups of users and librarians (respondents) are important. When extracting stereotypical attributes on social groups we need to take into account both generalised appreciations and the image of the individuals on the studied topic. Individual representation makes up a portrait by overlapping these attributes on the perception of librarians by the public. Direct approaches "bring up a note of realism, admitting that the process of categorising can be carried out by comparing new examples with examples already stored in one's memory" (CERNAT, 2005: 32).

7. IDENTIFYING THE STEREOTYPICAL PORTRAIT OF THE LIBRARIAN IN FICTION

Methods specific to fiction identified the librarian as "literary character" and the library as his environment. The stereotypes we have identified above through the methods of sociological research can be found in fiction as well, but with a surplus of artistry. We suggest the following books from contemporary world literature:

- Umberto ECO, with his already famous The Name of the Rose;

- Alberto MANGUEL, The Library of the Night and History of Reading;

- Vicki MYRON DEWEY, The Kitten of the Library of a Small Town Conquers the World;

- Mikkel BIRKEGAARD, The Library of the Shadows;

- Jacques BONNET, Des Bibliothèques pleines de fantômes [On libraries Full of Phantoms], Paris, Denoël, 2008.;

- Marc BARATIN, Christian JACOB (eds.), The Power of Libraries;

7. Identifying the stereotypical portrait of the librarian in fiction

- Carlo FRABETTI, The Hell-Book;

- Jorge Luis BORGES, Aleph and The Sand Book;

- Rui ZINK, The Reader in the Cave; x Emmanuel DELHOMME, Un librair en collere [A Librarian in Anger], Paris, L'Editeur, 2011.

- In Romanian fiction, we emphasise the following books:

- Mihail VAKULOVSKI (2012). Bibliodioteca (Istorie) [Biblidiotheque (History)]. București: Casa de pariuri literare ;

- George ARION (2014). Insula cărților [Book Island]. București : Crime Scene Press;

- PAN IZVERNA (1990). Bătrânul anticar [The Old Second Hand Book Seller]. București: Cartea Românească.

In fiction, when it is about the space of libraries, writers are almost all fascinated by mysteries, by the secrets of the history, by the culture hidden here, and unveiled only for the initiated. Moreover, the "inert" traditional library is in many cases the scene of intertwined detective stories. The librarian characters have in common the fact that they are males with good attributes: refined, charming, intelligent, visionary intellectuals that are true heroes with higher abilities of interpreting the reality by deciphering the secrets of the old books hidden in libraries. Specialists estimate that "the spreading and intensity [of stereotypes] knows important historical fluctuations" (CERNAT, 2005: 125): this explains why the model of a male librarian is so strong in fiction reflecting yet another professional reality deeply rooted in the past – the myth of the male librarian.

We have come across the attributes mystical, enigmatic in the book The Old Second Hand Book Seller by Romanian author Pan Izverna (1990). In the book, the old second hand book seller – a pair of librarians – is described wearing a beard and white hair, glasses and clothes from old times, and he looks like a magus, a preserver of the secrets in the books.

These features are also found in world literature. Haphazardly or not, successful books about or whose action has a library as a background, include in their titles and content references to a 'sanctuarylibrary" evoking an inhibiting sub-conscious fear of those who wish to get closer to the "memory of humankind" such as a library, to the books and overwhelming knowledge within this space. When we have to deal with descriptions of libraries and/or librarians, Romanian literature reflects the bitterness of a routine space, of professional blockage, of material lack (Biblidiotheque and The Island of the Books) where authentic intellectuals cannot find themselves and librarians are superficial, blasé, even hostile to the public. There are both negative and positive attributes in the responses of the participants to the focus-groups that meet the attributes described in fictions.

As for the anecdotes, more than other sources, professional stereotypical attributes related to librarians are pejorative: organised, obsessed with returning the books and with the book fund's integrity, while the library is a place where silence is mandatory. Librarians are portrayed with the help of stereotypes and seen as a reference model.

A blond young woman enters a library and says loudly at the reception desk: - Hi, I would like a hamburger and a coca-cola... - Muss, says the librarian, this is a library... Then the blonde-haired person whispers: - Sorry, could I also get some fries?	Five surgeons, ex-mates in college, meet at a symposium some years after graduation. In the evening, they go out for a beer and start talking about surgery. The first one says, "I love accountants: when you open one up, everything inside is counted and calculated". The second one says, "I prefer electricians – when you open one up, everything is color-coded...there is no way you mix up things".
A person goes to a library and asks for a book on suicide. The librarian refuses saying: - No way, you won't bring it back!	The third one says, "I like librarians – inside them everything is put in alphabetical order...it's easy to work".

8. CONCLUSIONS

The formation of stereotypes is strongly dependent on social factors: this is why libraries – even Romanian ones – are looking for solutions meant

8. Conclusions

to re-brand institutionally and professionally. Libraries change in architecture, functionality, equipment with new communication technologies, and concepts related to receiving and guiding the readers. The spaces become more welcoming and lightened. An anecdote on the topic of libraries says that a traditional librarian would sacrifice a reader for a book, while a last generation librarian would sacrifice a book for a reader – which shows that the focus changes on the users without, of course, neglecting the integrity of a collection and the rigours of a library management. The new library still has the cult for the undoubted value of the word, for information (assimilated by the metamorphoses and the avatars of contemporary reading), but it is moving with the users being, at the same time, a space for intercultural dialogue where there is an answer for each question.

Though stereotypes are difficult to dismember through cultural education, through good promotion, they can be changes with the help of the different mass media. Another way of changing the preconceived ideas would be through information culture taught as a subject in universities and high schools. Libraries of all kinds strive to diversify their services and to update them depending on the demands of the public, and also to modernise their spaces; the greatest issue here is the lack of funds, of trained staff and motivation in the staff, the need for special management equipment and programmes, the updating of the collections, the difficulty of persuading the accountants of the institutions coordinating libraries that we need to update libraries depending on the level of the "knowledge society" we live in – which could contribute to the improvement of the Romanian info-documentary institutions and, hence, to the quality of the educational process itself

NOTE

† There is a generous and suggestive gallery of images portraying librarian stereotypes all over the world at: Images for librarian stereotype: https://www.google.ro/search?q=librarian+stereotype+pictures&es_sm =93&tbm=isch&tbo=u&source=univ&sa=X&ei=RGAU6GCEeb07Aa7to GgDQ&ved=0CCUQsAQ&biw=1280&bih=644.

REFERENCES

1. Adams, Annis Lee, Emanuel, Jenny et al. (2007). Rebranding the Librarian Profession. HLAA Annual Conference, November 10.

2. Babbie, Earl (2000). Practica cercetării sociale. Iaşi : Editura Polirom.

3. Bourhis, Richard Y., Leyens, Jacques-Philippe (1997). Stereotipuri, discriminare şi relaţii intergrupuri. Iaşi : Polirom.

4. Cernat, Vasile (2005). Psihologia stereotipurilor. Iaşi : Polirom.

5. Grawitz, Madeleine (1996), Méthodes des sciences sociales, 10-ème éd. Paris : Dalloz.

6. Herscheberg-Pierrot, Anne (1997). Stéréotypes et clichés. Paris : Nathan, 1997.

7. Images for librarian stereotype: https:// www.google.ro/ search?q= librarian +stereotype +pictures &es_sm=93&tbm= isch&tbo= u& source=univ&sa=X&ei=RGAU6GCEeb07Aa7toGgDQ&ved=0CCU QsAQ&biw=1280&bih=644.

8. Moscovici, Serge, Buschini, Fabrice (2007). Metodologia ştiinţelor socio-umane. Iaşi : Polirom.

9. Samson, Mariana (2011). Elemente de istorie a noțiunilor de clișeu, poncif, stereotip. Câmpulung Muscel : Larisa.

10. Silverman, David (2003). Interpretarea datelor calitative. Iași: Polirom.

CHAPTER 3

Strategic Management Model for Academic Libraries

Khalfan Zahran Al Hijji

Department of Information Studies, Sultan Qaboos University

ABSTRACT

This study utilized the qualitative approach including content analysis of literature and in-depth interviews in order to design Strategic Management Model for Academic Libraries. The literature reviewed indicates many models for strategic management, which were generally made to suit organizations from different sectors. These models were found lack important elements that academic libraries need to connect their strategies to the overall visions and goals of their parent institutions. The Model, which is presented in the present paper therefore, attempts to bridge this gab by providing a new route for articulating and implementing strategies for academic libraries through three main stages: pre-planning stage; planning stage; and post- planning stage. The first stage starts by providing the planning team or committee with skills and knowledge of strategic management. The second stage is achieved through the fulfilment of two components of the Model: strategy formulation; and strategy implementation. The last stage however, concerns with the evaluation process of the strategy to ensure that the quality of services provided, and the performance of all library units and employees are compatible with vision and objectives of the library, and aligned with the overall goals of the Mother Institution.

Comments: This paper is a part of my PhD thesis titled "Strategic Management and planning practices at Academic Libraries in Oman", which was submitted to the University of Sheffield, UK. The Value of Model presented in this paper would be seen in new issues that are involved in its' structure such as pre planning stage and the alignment with the parent institution strategy, which are not included clearly in other Models. Moreover, all models that the Author could reach through the literature review of library management, either describes strategic planning steps, or strategy perspectives, and none has included the whole process and components of strategic management, which was the focus of this study. The Model also contains three components which have not been seen in other models.

KEYWORDS

Strategic Management Model; Strategic Management; Strategic Planning; Academic Libraries; Library Management; Performance Measurement; Quality Control

1. INTRODUCTION

Strategic Management (SM) is an ongoing process concerned with the identification of strategic goals, vision, mission and objectives of an organization along with an analysis of its current situation, develop appropriate strategies, put these strategies into action, and evaluate, modify or change these strategies when needed (G. Dess, G. Lumpkin, A. Eisner 2008). In order to help leaders and decision makers to adopt optimal strategies at their organizations, strategists and researchers have introduced models for different aspects of SM (G. Bowman, D. Asch 1993, L. Byars, L. Rue, S. Zahra 1996, M. Coulter 2005, J. Thompson, F. Martin 2005, B. De Wit, R. Meyer 2007). Although these models mostly include components of SM process as seen in the literature, they lack some important features that are required in academic libraries such: "Communication" and "Alignment", as very important features to connect the library's strategy to goals and objectives of parents' institution. This paper aims to present a Model for the SM process in the academic libraries, as an attempt to help academic librarians: to adopt SM principles and practices in their libraries; to overcome the obstacles that they might face in their future strategies; and to provide a guideline for formulating and implementing strategies as well as for the performance measurement process.

2. METHODOLOGY

Content analysis of literature provided valuable information about previous models and concepts of strategic management. More important data were accorded through interviews with leaders of academic libraries in Oman, an Arabic country located in the south east of Arabian Peninsula. Results of the two methods asserted the importance of designing SM model for academic libraries reflects the interrelations between the libraries and their parent institutions.

3. RELEVANT WORKS

The literature indicates many models for SM and strategic planning. Some of these were generally made to suit organizations from different sectors and includes phases of the SM process (L. Byars, L. Rue, S. Zahra

1996, M. Coulter 2005). Others were directed for nonprofit organizations such as the Bryson's framework for public and nonprofit organizations (J. Bryson 1988), which was applied at Indiana University Bloomington and was seen by McClamroch et al (2001) worthwhile for its' Libraries. Some others emphasize certain features of the strategy as seen in the Conceptual framework of "Strategy as Moral Philosophy" which was developed by Singer (1994). For academic libraries, Birdsall and Hensley (1994) developed a strategic Planning Model for Academic libraries composed of six components: selecting the most appropriate members of an organization and selecting an organizational structure to conduct planning; scanning the environment, which begins with a review of organizations mission and vision, followed by a review to major impact analysis of strengths weaknesses opportunities and threats; analyzing strategic options; designing unit plans; accepting the agenda, which involves affirming existing goals and develop new ones if necessary; and lastly adopting the strategic plan. Frameworks also were seen concerning: aspects of SM in libraries such as the "Mixed-model CAF-BSC-AHP and PAQ-SIBi-USP", which was developed by Melo and Sampaio (2006) for measuring the quality of academic libraries and information services; or facilities of libraries such as the Hofmann's strategic planning framework for information technologies for libraries.

4. STRUCTURE OF THE MODEL

SM is achieved through three basic phases: strategy formulation, which is concerned with defining the mission, vision and goals, and concluded by selecting a suitable strategy; strategy implementation, which involves selecting leadership and an organizational structure, and providing the required resources; and lastly, strategy evaluation and control, which includes the adoption of measurement tools for evaluating the performance of the organization.

4. Structure of the Model

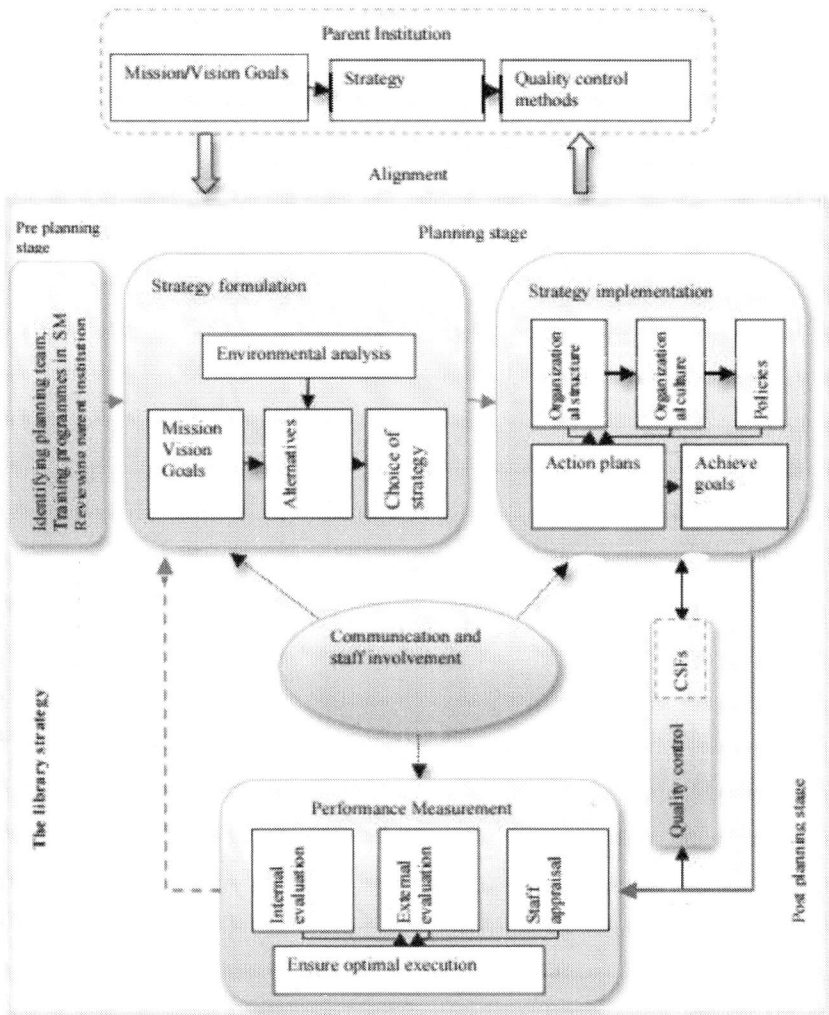

Figure 1 strategic management process Model for academic libraries

This Model as shown in Figure (1) is constructed to be congruent with these three stages. However, it includes other aspects that are especially designed to suit academic libraries. All the components of the model are merged into the following three stages: pre-planning stage; planning stage; and post-planning stage. Dealing with these issues successfully will enable the organization to reach its goals and to fulfill its vision. To ensure the optimal development and implementation of the SM stages, there are three connection channels linking the components of the model into a coherent whole. These are: an alignment channel, which ensures the total consistency of the library's mission, vision, and strategy with those of the parent institution; quality control, which links the implementation of the strategy to the performance measurement process; and a communication and involvement channel, which builds the engagement of employees from different management levels into all steps of the SM process.

5. PRE PLANNING STAGE

The aim of this stage is to prepare for strategic planning. It includes three activities: identifying the planning team or committee; providing the team with skills and knowledge of strategic management; and conducting a review of the parent institution's strategy. The first and the second activities are the responsibility of library leaders who are required to select highly skilled professionals to form the team, and then to support them with the training programs according to their needs. However, the third activity and most steps of the following stages will be carried out by the appointed team. Members of the planning team should be selected according to two criteria: firstly, they must be all professionals holding degrees in library science or management; and secondly, they must represent all units and management levels in the library. Sub committees may also be required to facilitate the optimal communication between the main team and stakeholders of the library in order to ensure the continuing flow of information from different units of the library to the team. Despite the selection of highly experienced employees to compose the team, some or maybe all of them might not have the basic skills and knowledge of the SM process. Therefore, and in order to ensure the right direction in formulating and implementing strategies, training workshops in this regard should be designed to provide the team members with the required knowledge.

After the completion of training, all members of the team will be supposed to have the ability to think strategically, and then to identify the elements of strategic planning and the process of strategic management. From this point, the team will start its duty by reviewing the strategies and other kinds of plans of the parent institution. The present model supposes that the parent institution already has its own strategy when leaders of the library commence the work in strategic planning for information services. Thus, it is essential for the planning team to visit and review the university's strategy and analyze its components, in order to identify the basic elements that must be taken into consideration for the library's strategy, and which ensure the successful alignment between the two strategies. However, if the university has not yet developed its strategy, the library planning team should consider, at least, the mission and vision statements of the university if they are available, or meet university principals to clarify their vision for the University.

5.1. Planning stage

This stage is achieved through the fulfillment of two components of the Model: strategy formulation; and strategy implementation. The first one includes four steps: identifying vision, mission and goals of the library; conducting an environmental analysis; specifying alternatives; and choosing the strategy. The second component also aims to achieve the strategic goals of the library through four steps: developing an organizational structure; identifying the organizational culture; developing policies; and identifying action and operational plans.

5.2. Strategy formulation

Leaders and principals of all organizations should articulate: a vision shaping the direction and future of their organizations; a mission statement explaining what exactly the organization does, and justifying the reason for its existence; and goals that are targeted to be achieved in order to achieve the vision. Before commencing the formulation of library strategy, the planning team should be sure of the existence of these three important components of SM in the library. If this is the case, the next step is to ascertain that these components are conforming to good SM practice, as described in the literature, and then to ensure their alignment with those of the parent institution. This Model therefore, suggests that library principals and planning team identify in an earlier stage of formulating the strategy, the purpose of the library and the type of information services it offers, and then articulate vision and mission statements, and library goals accordingly. The most important thing in this step is to ensure the participation of all library leaders in articulating these issues, and the commitment of all staff to achieving them.

Once vision and mission statements are articulated and shared, and goals and objectives are developed, the environmental analysis should be carried out. In this process, all strengths and weaknesses of the library, and all opportunities and threats that may help or affect the implementation of its strategy are brought together for analysis and discussion. For this reason, and as a result of its paramount importance for developing strategies, the environmental analysis has been included in the heart of the first component (strategy formulation) of the present Model. The planning team should work very closely with librarians from different departments and from different management levels to identify and analyze these issues. Having compared strengths to weaknesses and opportunities to threats will enable the team to address alternatives that may help to overcome weaknesses and threats and enhance strengths and opportunities; and then by comparing these alternatives, the best one will be chosen as a strategy.

6. STRATEGY IMPLEMENTATION

When the strategy is formulated, the next step is to put it into action through the identification of human resources, the allocation of budgets, and the adoption of efficient procedures. To achieve this successfully, the Model suggests that the planning team should work in developing the following issues. First, the organizational structure of the library, which facilitates the optimal communication between all employees, and specifies decisionmaking authority; the structure should be designed to match the strategy and to suit the size of the library. Second, the organizational culture, which includes basic assumptions and values that must be shared by employees as the way to perceive, think, feel, and behave in the library. When people learn to deal successfully with problems, this becomes a common language and background. Thus culture arises out of what has been successful for the organization (L. Aiman-Smith 2010). Al Hijji indicated that social values and behaviors of staff members and library users affect the performance of some libraries. Therefore, it is critical to identify, in advance, how to manage these problems and how employees will share cultural values that raise the level of commitment to the successful implementation of strategies. The third issue, which must be taken into account in the implementation process of strategies is the development of policies and procedures that are essential to be followed in providing services and managing different resources. The planning team, therefore, should review any policies and procedures that exist in the library in order to check their consistency with the requirements of the strategic plan and to adapt them accordingly. The final issue is the development of action or operational plans. These plans are intended to achieve each of the strategic goals of the library.

7. POST PLANNING STAGE

When the strategy is in place and steps towards its implementation have been taken, the next step is to ensure that the quality of services provided, and the performance of all library units and employees are compatible with its requirements. This should be done through close supervision of the employees, and continuous evaluation of services. Quality control is achieved through a system of quality audits at every stage of the process of implementing strategies. This system should be connected to a set of critical success factors. The measurement process then should be carried out to explore the extent to which the library performance meets these factors. The researcher is of the opinion that the combination of the two methods will help in getting better results. Thus, the Model suggests that internal evaluation should be conducted on an ongoing basis and applied by librarians at every level of the library. The external evaluation by researchers and academic bodies however, should be conducted at least once during the period of the strategy. Different kinds of evaluation tools such as surveys, interviews, and statistics should be taken into consideration. Library directors should use the results of different kinds of measurement to monitor the direction of the strategy implementation, and then make decisions to amend action plans accordingly. The measurement process also includes staff appraisal. People are the drivers of any organization. They develop and implement the strategies, and they measure the success of the strategies. Thus, the measurement of libraries should include an ongoing appraisal of staff to assess their ability to perform their duties, and consequently to ensure they are working towards achieving the strategic goals of the library. For this reason, special forms should be designed reflecting the nature of the different job of libraries, and connected to the strategy of the library. The annual basis of appraising staff is seen as reasonable by the researcher. However, the appraisal should be made against specific levels of performance and objectives that are targeted by action plans. Moreover, the evaluation should explore the strengths and the weaknesses of each staff member, and should be connected to the motivation system and training programs. The results of the different kinds of measures help the planning team and librarians to identify issues that must be taken into consideration in developing further strategies.

8. COMMUNICATION AND STAFF INVOLVEMENT

Library leaders and the planning team should work together to develop an effective system for ensuring the commitment of all library staff to strategies of their libraries. This requires forming different kinds of committees, which aim to improve co-operation and understanding between management and staff, and to ensure the participation of employees in decision-making and planning activities, which consequently make them feel that they are role players in the management process of their libraries. Moreover, efficient communication should be established between librarians and staff members of other units of the university to ensure that information services in the library meet the needs of their units. Furthermore, good communication with the university management will provide library leaders with the opportunity to explain the problems that they face in implementing strategies to the university principals in order to get their support to solve them. For all these reasons communication and staff involvement is located at the heart of the present Model.

9. CONCLUSION

The Model, which is presented in this paper is a result of a comprehensive survey of SM literature. It provides a guideline for academic libraries' leaders to adopt SM principles and practices in their libraries and to overcome the obstacles that they might face in their future strategies. The Model comprises the basic components of SM: strategy formulation; strategy implementation; and strategy evaluation and control, and it is applied through three successive stages: pre-planning; planning; and post planning. To ensure the optimal development and implementation of these stages, the Model offers three connection channels linking the components of the model into a coherent whole. These are alignment, quality control, and a communication and involvement channels.

ACKNOWLEDGEMENT

The author would like to thank all the anonymous participants who have kindly taken part in this study, and have given up valuable knowledge through long interviews.

REFERENCES

1. Singer: Strategy and Moral philosophy. Strategic Management Journal vol 15 no 3 . (1994). p 191-213.

2. De Wit and R. Meyer: Strategy synthesis: resolving strategy paradoxes to create competitive advantage. Andover, Cengage Learning (2007).

3. Birdsall and O. Hensley: A New Strategic Planning Model for Academic Libraries. College & Research Libraries vol 55 no 2 (1994). p 149–159.

4. G. Bowman and D. Asch: Strategy management. London: Macmillan (1993).

References

5. G. Dess, G. Lumpkin and A. Eisner: Strategic management: Texts and cases. 4th ed. Boston: McGraw-Hill (2008).

6. J. Bryson: A Strategic Planning Process for Public and Non-profit Organizations.

7. J. McClamroch, J. J. Bryd and S. L. Sowell: Strategic Planning: Politics, Leadership, and Learning. Journal of Academic Librarianship, vol 27 no 5 (2001), p 372-378.

8. J. Thompson and F. Martin: Strategic management: Awareness and change. 5th ed. Australia: Thomson (2005).

9. K. Al Hijji: Strategic management and planning practices in academic libraries in Oman. PhD, University of Sheffield, UK.

10. L. Melo, and M. Sampaio: Quality Measures for Academic Libraries and Information Services: Two Implementation Initiatives – Mixed-model CAF-BSC-AHP and PAQ-SIBi-USP (2006), [available online] http://iat14290s1.chopin.2day.com/doclibrary/public/Conf_Proceedings/2006/MeloSampaiopaper.pdf [Accessed 10/05/2013].

11. L. Aiman-Smith: What Do We Know about Developing and Sustaining a Culture of Innovation: Organizational culture [available online] http://cims.ncsu.edu/downloads/Research/71_WDWK_culture.pdf [Accessed 21/06/2010].

12. L. Byars, L. Rue and S. Zahra: Strategic management. Chicago: IRWIN (1996).

13. L. Byars, L. Rue and S. Zahra: Strategic management. Chicago: IRWIN (1996).

14. Long Range Planning, Vol 21 no 1 (1988), p. 73-81

15. M. Coulter: Strategy management in action. New Jersey: Person education International (2005).

16. M. Coulter: Strategy management in action. New Jersey: Person education International (2005).

17. U. Hofmann: Developing a strategic planning framework for information technologies for libraries. Library Management Vol. 16 no 2, p 4 –14.

CHAPTER 4

Factors That Increase The Probability Of A Successful Academic Library Job Search

Max Eckard[1], Ashley Rosener[2], Lindy Scripps-Hoekstra[3]

[1]*Metadata & Digital Curation Librarian, University Libraries, Grand Valley State University, Allendale, MI 49401, USA*

[2]*Liaison Librarian in Professional Programs, University Libraries, Grand Valley State University, Grand Rapids, MI 49504, USA*

[3]*Liaison Librarian in Liberal Arts Programs, University Libraries, Grand Valley State University, Allendale, MI 49401, USA*

ABSTRACT

Finding a position in an academic library can be challenging for recent Library and Information Science (LIS) graduates. While LIS students are often encouraged to seek out experience, network, and improve upon their technology skills in hopes of better improving their odds in the job market, little research exists to support this anecdotal advice. This study quantifies the academic and work experiences of recent LIS graduates in order to provide a better understanding of what factors most significantly influence the outcome of their academic library job searches. The survey results demonstrate that the job outlook is most positive for candidates who applied early, obtained academic library experience (preferably employment), participated in professional conferences, and gained familiarity with committee work.

KEYWORDS

Job hunting; Academic libraries; Education

1. INTRODUCTION

Finding a position in an academic library can be challenging for the newly minted library and information science (LIS) graduate. Fortunately, there is no shortage of job-seeking advice for prospective candidates. From librarian blogs to professional magazines and websites such as "I Need a Library Job" (www.inalj.com), suggestions and recommendations abound. In their 2010 C&RL News piece, Cannady and Newton recommended making the "best of the worst of times" (p. 210) and gave specific advice for each phase of the job search: application, phone interview, and face-to-face interview. Singleton (2003) recommended building a network of professional contacts and being prepared to show effectiveness in professional assignment, professional development, and service. In "Academic Library Job Search Blues," Baker (2010)advised making oneself more marketable by finding internships, becoming involved in organizations, and taking a balanced blend of classes.

While this advice may be sound, it is largely based upon the subjective experiences of the authors. The overabundance of such anecdotal advice can make a difficult job search even more challenging. With the variety of choices in coursework, internships, and other opportunities students are advised to pursue during their short time in graduate school, how will they know which activities are the most beneficial for ensuring a successful academic library job search? The objective of this study is to quantify the experiences of recent LIS graduates to better understand what factors might help job seekers obtain their first professional position. By comparing the academic library job search outcomes of survey respondents, this study shows that applying early, obtaining academic library experience (preferably employment), participating in professional conferences, and gaining familiarity with committee work increases the probability of being able to obtain a first post-degree position. The results of this study also identify trends in LIS students' graduate school involvement (academic and extracurricular) and recent LIS graduates' perspectives of the academic library job market and the process of securing a job.

2. LITERATURE REVIEW

The labor market for LIS graduates has been described, in general, as "relatively 'recession proof'" (Morgan & Morgan, 2009, p. 299), and indeed, librarianship has endured the Great Recession of the late 2000s. The 2012–13 edition of the Occupational Outlook Handbook stated that employment of librarians is actually expected to grow by 7% from 2010 to 2020 (U.S. Department of Labor & Bureau of Labor Statistics, 2012, 2013). While slower than average for all occupations, this is growth nonetheless. According to an analysis done in 2012, salaries are up 5% while unemployment for recent graduates held at 6%, suggesting that positions are available, with similar numbers for 2013 (Maatta, 2012 and Maatta, 2013).

2.1. Academic Library Job Market

According to the Bureau of Labor Statistics, later in the decade prospects should be even better as older library workers retire and population growth generates openings (U.S. Department of Labor & Bureau of Labor Statistics, 2012, 2013). However, it should be noted that this optimistic prediction is reminiscent of the early 2000s expectation that the availability of academic library jobs would increase "due in large part to the 'graying' of the profession" (p. 408), an estimation that never came to fruition because budget constraints, especially in local government and educational services, slowed demand for librarians (Tewell, 2012).

The academic library job market for recent LIS graduates is competitive, and for those lacking significant practical experience, it is a "potentially insurmountable challenge" (p. 422) (Tewell, 2012). Nearly three-quarters of academic librarian positions in 2011 preferred or required work experience (Triumph & Beile, 2011). Also, in 2012 almost three-quarters of academic library jobs were non-entry level, and of those 26% of all job advertisements were administrative, creating a tough market for recent graduates (Tewell, 2012).

2.2. Search Committee Perspectives

The academic library job search and hiring process provides an additional challenge for recent graduates. During times of economic distress when employment opportunities are few and the number of applicants increases, candidates must be aware of the rigorous academic library job search process (Durán, Garcia, & Houdyshell, 2009). In their survey of search committees, Hodge and Spoor found that 78% of respondents are receiving more applications per position opening now than in previous years, but 80% do not interview more candidates (2012). Search committee members can suffer from fatigue while reviewing such

2. Literature Review

large numbers of applications and may overlook qualified candidates (Howze, 2008).

The Association of College and Research Libraries' Discussion Group of Personnel Officers "agreed that previous library experience was an important requisite for an entry level position. The majority also indicated that the experience should be in an academic library" (Neely, 2011, p. 4). Previous work experience is an indicator of future job performance, as Wheeler, Johnson and Manion highlighted when they suggested that questions about a candidate's experience are more effective than situational questions(2008). Demonstrated performance of job requirements is very important. In their 2010 study, Wang and Guarria found that 90% of the 243 survey respondents (individuals who served on faculty search committees) believed that a demonstrated ability to perform job requirements was very to extremely important (p. 83).

In addition to work experience and the ability to perform job requirements, candidates must demonstrate they are leaders, not just workers. Search committees want librarians who are creative, proactive, risk takers, innovators, independent yet collaborative, lifelong learners, and visionaries (Harralson, 2001). Reeves and Hahn (2010) reminded LIS students that employers prefer individuals who have good communication skills, work well with others, take initiative, are adaptable and dependable, and have a "service orientation, a predilection for collaboration and cooperation [and] a penchant for participating in teams" (p. 118). Employers also look for applicants who can acclimate quickly to organizational culture. In Wang and Guarria's survey, over 90% of the 243 survey respondents said potential fit is very or extremely important in an academic library(2010). For recent graduates with less work experience, this further emphasizes the importance of potential fit with the organizational culture.

3. METHODOLOGY

In order to learn more about the graduate school experiences and job search successes of recent LIS graduates, an electronic survey was created using SurveyMonkey (Appendix A). In March 2013, a link to the survey was emailed to 2008–2012 graduates from the LIS programs at the University of Illinois at Urbana-Champaign and North Carolina Central University. The survey link was also emailed to members of the ALA New Members' Round Table (NMRT) listserv, distributed on index cards to ACRL 2013 conference attendees during a related poster presentation, and electronically posted on the ACRL New Member Discussion board. At the request of the LIS program at Dominican University, the survey link was not emailed to their graduates until July 2013. In addition to being the three universities from which the three researchers graduated, these institutions represent a cross section of different LIS programs around the United States in that they have different U.S. News & World Report rankings, program formats, enrollment numbers, and local job markets. Respondents from the three universities and the NMRT listserv provided a diverse sample of recent LIS graduates from ALA-accredited programs around North America. While the respondents graduated in different years throughout the last decade, the survey questions only addressed their time in graduate school.

The survey questions were divided into seven primary categories: basic information, job search, professional effectiveness, professional development, service, technological competency, and previous careers. Questions focused on the student's graduate program, the parameters of their job search, their academic and work experience, as well as other skills or professional involvement that might impact their success in landing a first job. In addition to these questions, the survey concluded with an open-ended response question that asked, "What advice do you have for current LIS students?" This question and all other free-response questions were coded by the researchers and differences in coding were resolved by mutual agreement. The inter-coder reliability averaged

Kappa = .9, which represents an almost perfect agreement according to Landis and Koch (1977, p. 165).

3.1. Limitations

The results are based upon a limited sample size of recent LIS graduates since the survey was sent to three specific universities. This limitation was offset to a certain extent by also sending the survey to the NMRT listserv, posting it on the ACRL New Member Discussion board, and passing out links at the 2013 ACRL Conference, which allowed for gathering responses from graduates from many other LIS schools. Due to the numerous avenues by which the survey link was dispersed, it is not possible to calculate the response rate. The results are also limited due to the fact that survey questions were optional and not all elicited an adequate number of responses for the following analysis.

4. RESULTS

Out of 360 total survey respondents, 56% (N = 201) indicated that they wanted to work in academic libraries. Eighty-two respondents wanted to work in public libraries with smaller numbers selecting "special libraries" or "other". The survey respondents provide a varied sample of LIS graduates. The 201 respondents who selected academic librarianship were recent LIS graduates from the years 2005–2013 with the highest percentage (28%) graduating in 2011. Respondents represented 33 different LIS programs with the highest number of students graduating from the University of Illinois (56) and Dominican University (39).

4.1. Successful Respondents

Successful respondents are defined as the subset of survey respondents who were able to find a full-time or part-time, tenure-track or professional academic library job after graduation. Respondents were asked about the outcome of their job search and 186 (n = 186) individuals responded to this question. Sixty-eight percent (126) indicated that their job search was successful and 23% (60) reported an unsuccessful job search. In the following analysis, percentages are based upon the number of successful respondents who answered each question. As this study's aim is to determine the factors that lead to success in obtaining a job, our analysis focuses only on the successful respondents.

4.2. Basic information

Over 60% (75) of our successful respondents were full-time students. In terms of program format, 44% (56) attended their program in-person, 30% (43) enrolled in a hybrid program, and 24% (30) completed their degree online.

4.3. Job search

Respondents were asked when they began applying for jobs, and 62% (75) reported having started the application process four months or more prior to graduation. In terms of the number of positions applied for, the highest percentage (32%) of successful respondents applied for five or fewer jobs, while the lowest percentage (8%) of respondents reported applying to over 50 positions (Fig. 1). Almost half of successful respondents (52) found a position before graduating, and 40 respondents found a position within one to three months after graduation.

4. Results

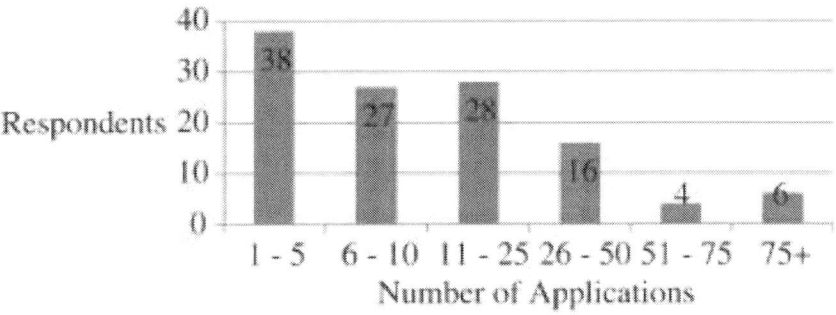

Figure 1. Number of successful applications.

Just over 50% of respondents (65) were limited to a specific geographic area with 56 reporting the ability to move for a position. Almost half found a job as a reference and instruction librarian (Fig. 2). The next most common type of position was one in technical services.

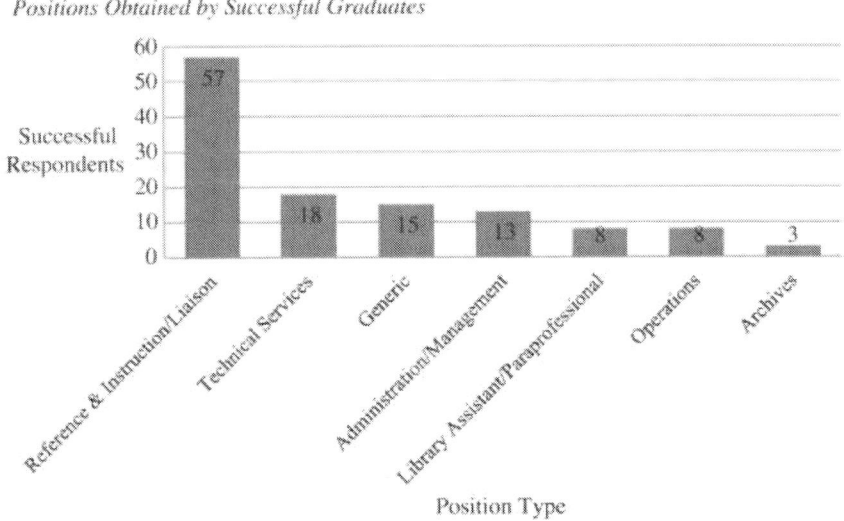

Figure 2. Positions obtained by successful graduates.

The types of positions reported varied in status, and 86% (108) reported landing full-time positions. Over half of successful respondents (62) described their position as "professional," and a third (42) described their position as "faculty" or "tenure-track".

4.4. Professional effectiveness

Survey respondents were also asked to select their coursework choices and extracurricular activities from a provided list (respondents could select multiple options). Results indicated that 82% (103) took at least one class in academic librarianship, 66% (83) worked in an academic library, and approximately 51% (64) had an internship in an academic library. Some respondents also gained experience by volunteering at libraries. Of these respondents, 23% (29) volunteered in a public library, 15% (19) in a special library, and 8% (10) in an academic library.

4.5. Professional development

Approximately 67% of respondents reported having attended conferences (84) during graduate school. A similar percentage attended workshops and seminars (77). Almost 25% (30) participated in conferences. Finally, 14% (18) had either authored or co-authored a publication of some type.

4.6. Service

Over 50% (81) of the respondents were involved in library associations and/or were active in library student groups (65).

4.7. Technological competency

In response to a general question about participation in elective technology courses while in graduate school and self-reported technological competency level, seventy-six percent (96) of respondents rated themselves as either "Competent" or "Very competent" with technology.

4.8. Previous careers

Just over 50% (64) came to librarianship as a second career and 86% (75) of them worked for ten years or less in their previous careers. Of these respondents, 66% (58) felt their previous job helped them land their library job. Respondents were asked to name their previous career(s), and their answers were coded based upon the occupational groups identified by the Bureau of Labor Statistics' in the Occupational Outlook Handbook. For successful respondents, the most common previous careers were in the fields of education, training, or libraries.

4.9. Respondent advice

The survey concluded with an open-ended response question crafted to elicit perceptions on how to be successful in securing a job. Respondents were asked what advice they had for current LIS graduate students (Fig. 3). Answers varied, but some pieces of advice came up repeatedly. Of successful respondents, 40% felt that experience was important and 18% recommended networking. Twelve percent recommended job market preparation, perhaps in the form of help with formatting and writing cover letters and resumes. Just 11% pushed for graduates to increase their technological savvy, and only four percent respondents recommended that students be willing to move for a job.

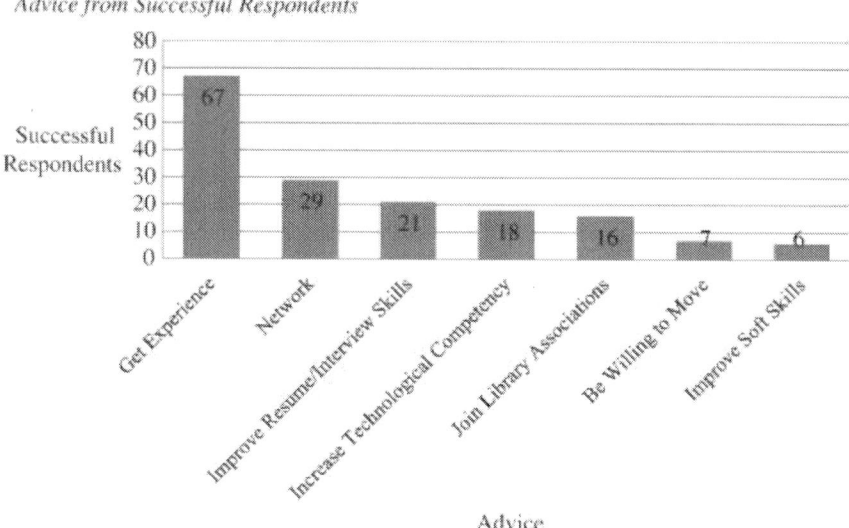

Figure 3. Advice from successful respondents.

5. DISCUSSION

We compared successful and unsuccessful job seekers in order to determine whether any trends existed for either group. Overall, the two groups were largely similar. Of the seven question categories, only certain factors in job search, professional effectiveness, professional development, and service made a significant difference in improving a candidate's odds of success in landing a job. Among these factors were academic library experience, committee work, conference attendance, and publications. These factors significantly increased or decreased the odds of getting a job after graduation; odds ratios were calculated and analyzed for respondents using 95% confidence intervals (Ekstrom & Sorenson, 2011, p. 321).

5. Discussion

5.1. Job Search

Recent graduates were asked when they began applying for jobs. The odds ratio (6.78) showed that candidates who began applying for jobs four to six months before graduation were nearly seven times ($p = .0092$) more likely to obtain a job than candidates who did not. Based on the confidence intervals (1.26, 5.6), we can be 95% confident that the entire population of LIS graduates who begin applying for jobs four to six months before graduation are between 1.26 and 5.6 times more likely to get a job compared to those who do not. Applying for jobs four to six months before graduation increased a candidate's odds of success more than any other factor.

5.2. Professional Effectiveness

Having any academic library experience increased the odds that a job seeker would be successful, although not all types of experience were significant. The odds ratio of getting a job for candidates who had been employed in academic libraries compared to those who had not was 4.71 ($p < .0001$). The 95% confidence interval for this odds ratio is between 2.41 and 9.22. Participation in an academic library internship or practicum also significantly improved the odds that a recent graduate would successfully find a job. Candidates who had participated in an internship or practicum improved their odds of success by 2.75 ($p = .0027$) compared to candidates who had not completed an internship or practicum, with the 95% confidence limits for the entire population of recent graduates ranging from 1.41 to 5.38.

5.3. Professional Development

Recent graduates were also asked about professional development experiences. Candidates who had attended conferences increased their odds of success by 3.33 ($p = .0002$) when compared to candidates without this experience, with the 95% confidence limits for the entire population being between 1.76 and 6.33. Candidates who had attended workshops and seminars increased their odds by 2.05 ($p = .023$), with 95% confidence limits from 1.1 to 3.83. Finally, candidates who had publications increased their odds by 4.83 ($p = .0246$), with 95% confidence limits from 1.08 to 21.6.

5.4. Service

The last category that significantly impacted success was service. Recent graduates were asked which service opportunities they participated in while in graduate school including association membership, student groups, committee work, volunteering, and fundraising. Of these, only committee work was significant. Candidates who did committee work increased their odds by 3 ($p = .0274$), with 95% confidence limits between 1.09 and 8.23.

6. CONCLUSION

This study sought to provide a better understanding of which quantifiable factors can significantly influence an academic library job search. Based on these results, time of application, academic library experience, committee work, conference attendance, and authoring or co-authoring a publication appear to increase the odds of a recent graduate's ability to break into the field.

In addition to the factors that improve the odds of success in a job search, this study also found other factors addressed in the survey did not appear to significantly affect a graduate's chance of securing a job. Among these factors were: program format, graduation date, grade point average,

6. Conclusion

enrollment status, study abroad experience, independent study, and technological skills. It is difficult to quantify whether potential fit impacts success in landing a library position, but the literature suggests that a candidate's personality and potential fit with a library is a large consideration when hiring committees make their hiring decision. Due to the fact that successful and unsuccessful respondents were fairly similar, it could be deduced that potential fit plays an important role in the hiring process.

While students are ultimately responsible for preparing themselves for a job search, LIS programs can do more to assist them. Based on this research, we recommend LIS programs provide opportunities for students to obtain some form of academic library work experience. Considering the importance of early application, LIS advisors should also ensure that students are prepared for the unique considerations of the academic library job search process. LIS programs should also focus on connecting students to professional development and publication opportunities.

Our research represents an initial look into several factors in the academic library job search, yet many other variables warrant further exploration. More research could be conducted on what job market preparations are offered at different LIS programs and the potential impact of online portfolios and professional social networking. While this study focused on academic library job seekers, public and other types of libraries have yet to be examined. Further research could also explore the experiences of unsuccessful library school graduates in order to determine whether trends exist.

For recent LIS graduates hoping to obtain academic library positions, the outlook appears positive for candidates who begin applying for jobs four to six months before graduation, and who are able to procure academic library experience (preferably employment), become involved with professional conferences, and gain familiarity with committee work.

While these conclusions may not surprise many in the field, this study represents a quantitative understanding of oft-given job seeking advice. The advice given by 40% of successful respondents to get experience was sound; as one respondent stated, "Experience makes all the difference!"

ACKNOWLEDGMENTS

Special thanks to the Grand Valley State University Statistical Consulting Center.

APPENDIX A. SURVEY INSTRUMENT

Library Survey

*1. What type of library did you most want to work in?
- Academic
- Public
- School
- Special
- Other (please specify)

Survey Complete!

Thank you for participating, the information you provided has been valuable to our research.

2. When did you graduate with your MLS/MLIS?
- Winter (December)
- Spring (April-June)
- Summer (July-August)

3. In what year did you graduate with your MLS/MLIS?

4. Upon graduation, did you find a library job?
- Yes
- No

5. In what type of library did you find a job?
- Academic
- Public
- School
- Special
- Other (please specify)

Appendix A. Survey Instrument

6. Is the job full time or part time?
○ Full Time
○ Part Time

7. What is the classification of your position?
○ Professional
○ Faculty/Tenure Track
○ Para-professional
Other (please specify):

Job Search

8. When did you begin applying for jobs?
○ More than 7 Months Before Graduation
○ 4-6 Months Before Graduation
○ 2-3 Months Before Graduation
○ 1 Month Before Graduation
○ Less than 1 Month Before Graduation
○ After Graduation
○ Other (please specify)

9. How many library positions did you apply for?
○ 1-5
○ 6-10
○ 11-25
○ 26-50
○ 51-75
○ Over 75
○ Other (please specify)

10. What percentage of job applications would you ESTIMATE resulted in advances to the next level in the hiring process (phone/in-person interviews, etc.)?

11. How long after graduation did it take you to find a library job?
○ Before graduation
○ 1-3 Months
○ 4-6 Months
○ 7-9 Months
○ 10-12 Months
○ More than a year

12. What was the title of your first library job after graduation?

13. Were you limited to a specific geographic area when applying for library jobs?
○ Yes
○ No

14. Which school did you graduate from with your MLS/MLIS degree?
☐ University of Illinois Champaign-Urbana
☐ Dominican University
☐ North Carolina Central University
☐ Other (please specify)

15. Which type of degree did you obtain?
○ MLS
○ MLIS
○ MIS
○ Other (please specify)
_____

16. What was your graduating GPA?
_____

17. What was your enrollment status?
○ Full time student only
○ Part time student only
○ Some combination of full and part time student

18. What was the format of your program?
☐ In person
☐ Online
☐ Hybrid

19. In which of the following experiences did you participate in during graduate school (check all that apply)?

	Public Library	Academic Library	School Library	Special Library	Other	None
Coursework	☐	☐	☐	☐	☐	☐
Employment	☐	☐	☐	☐	☐	☐
Internship/Practicum	☐	☐	☐	☐	☐	☐
Volunteer	☐	☐	☐	☐	☐	☐
Study Abroad	☐	☐	☐	☐	☐	☐

Other (please specify)
_____

20. Please indicate whether you took multiple elective courses in technology while in graduate school (yes or no), and rate your technological competency level:

	Not very competent	Somewhat competent	Competent	Very Competent
Yes	○	○	○	○
No	○	○	○	○

21. Which of the following professional development activities did you participate in while in graduate school? (check all that apply)
☐ Conference Attendance
☐ Workshops and seminars
☐ Independent study
☐ Publication
☐ Conference Participation (papers, presentations and poster sessions)
☐ Grant Writing
☐ Additional degrees

Other (please specify)
_____

22. Which of the following service opportunities did you participate in while in graduate school? (check all that apply)
☐ Join associations
☐ Student groups
☐ Committee work
☐ Volunteer
☐ Fundraising

Other (please specify)
_____

23. Is librarianship your first CAREER?
○ Yes
○ No

Appendix A. Survey Instrument

Previous Careers

24. Please describe your previous career(s).

[text box]

25. How many years were you in your previous career(s) before you began library school?

○ 1-2
○ 3-5
○ 6-10
○ 11-20
○ Greater than 20

26. Do you feel that your previous career(s) helped you to land a library job?

○ Yes
○ No
○ Indifferent

27. Why I feel that your previous career(s) helped me to land a library job

[text box]

Final Thoughts

We will be using this information to determine the extent to which graduate experiences aid in landing a job after graduation in an academic library setting.

28. Are there any other work OR life experiences/information that you typically share with potential employers at any point in the application/interview process? Please elaborate below.

[text box]

29. What advice do you have for current LIS students?

[text box]

30. What advice do you have for hiring managers?

[text box]

Survey Complete!

REFERENCES

1. Baker, S. (2010). Academic library job search blues. Library Journal (Retrieved April 5, 2012 from http://www.libraryjournal.com/lj/communityacademiclibraries/886717- 265/academic_library_job_search_blues.html.csp)

2. Cannady, R., & Newton, D. (2010). Making the best of the worst of times: Global turmoil and landing your first library job. C&RL News, 71, 205–212.

3. Durán, K., Garcia, E. P., & Houdyshell, M. L. (2009). From the inside out and the outside in: The academic library interview process in a tight economy. C&RL News, 70, 216–219 (Retrieved from http://crln.acrl.org/content/70/4/216.full. pdf+html)

4. Ekstrom, C. T., & Sorenson, H. (2011). Introduction to statistical data analysis for the life sciences. Boca Raton: CRC Press.

5. Harralson, D.M. (2001). Recruitment in academic libraries. College & Undergraduate Libraries, 8(1), 41–74.

6. Hodge, M., & Spoor, N. (2012). Congratulations! You've landed an interview. New Library World, 113(3–4), 139–161 (Retreived from http://www.emeraldinsight.com/ journals.htm?articleid=17024629)

7. Howze, P. C. (2008). Search committee effectiveness in determining a finalist pool: A case study. The Journal of Academic Librarianship, 34, 340–353

8. Landis, J. R., & Koch, G. G. (1977). The measurement of observer agreement for categorical data. Biometrics, 33(1), 159–174.

9. Maatta, S. L. (2012). A job by any other name: LJ's placements & salaries survey 2012. Library Journal (Retrieved from http://lj.libraryjournal.com/2012/10/placementsand-salaries/2012-

REFERENCES

survey/a-job-by-any-other-name-ljs-placements-salaries-survey-2012/)

10. Maatta, S. L. (2013). Placements & salaries 2013: Salaries stay flat; specialties shift. Library Journal (Retrieved from http://lj.libraryjournal.com/2013/10/placements-andsalaries/2013-survey/salaries-stay-flat-specialties-shift/)

11. Morgan, C. H., & Morgan, J. C. (2009). The effects of entering the LIS workforce in a recession: North Carolina, 1964–2005. Library Trends, 58, 291–300. http://dx.doi.org/ 10.1353/lib.0.0079.

12. Neely, T. Y. (2011). How to stay afloat in the academic library job pool. Chicago, IL: American Library Association.

13. Reeves, R. K., & Hahn, T. B. (2010). Job advertisements for recent graduates: Advising, curriculum, and job-seeking implications. Journal of Education for Library and Information Science, 51, 103–119.

14. Singleton, B. (2003). Entering academic librarianship: Tips for library school students. C&RL News, 64, 84–86.

15. Tewell, E. C. (2012). Employment opportunities for new academic librarians: Assessing the availability of entry level jobs. Libraries and the Academy, 12, 407–423 (Retrieved from http://muse.jhu.edu/journals/portal_libraries_and_the_academy/v012/12.4. tewell.html)

16. Triumph, T., & Beile, P. (2011). The trending academic library job market: An analysis of job ads from 2011, with comparisons to 1996 and 1988 studies. (Retrieved from http://www.academia.edu/2029979/The_trending_academic_library_job_market_An_analysis_of_ads_from_2011_with_comparisons_to_1996_and_1988)

17. U.S. Department of Labor, & Bureau of Labor Statistics (2012). Librarians. In Occupational outlook handbook (2012–13) (Ed.),

(Retrieved from http://www.bls.gov/ooh/ education-training-and-library/librarians.htm)

18. Wang, Z., & Guarria, C. (2010). Unlocking the mystery: What academic library search committees look for in filling faculty positions. Technical Services Quarterly, 27, 66–86.

19. Wheeler, R. E., Johnson, N.P., & Manion, T. K. (2008). Choosing the top candidate: Best practices in academic law library hiring. Law Library Journal, 100, 117–135.

CHAPTER 5

Current practices in library/informatics instruction in academic libraries serving medical schools in the western United States: a three-phase action research study

Jonathan D Eldredge[1], Karen M Heskett[2], Terry Henner[3] and Josephine P Tan[4]

[1]*Health Sciences Library & Informatics Center and Department of Family & Community Medicine, University of New Mexico, MSC09 5100, Albuquerque, NM 87131-0001, USA.*
[2]*UC San Diego Biomedical Library, UC San Diego, 9500 Gilman Dr. 0699, La Jolla, CA 92093, USA.*
[3]*Savitt Medical Library, University of Nevada School of Medicine, Reno NV 89557, USA.*
[4]*UCSF Library and Center for Knowledge Management, UCSF, 530 Parnassus Avenue, San Francisco, CA 94143-0840, USA.*

ABSTRACT

Background

To conduct a systematic assessment of library and informatics training at accredited Western U.S. medical schools. To provide a structured description of core practices, detect trends through comparisons across institutions, and to identify innovative training approaches at the medical schools.

Methods

Action research study pursued through three phases. The first phase used inductive analysis on reported library and informatics skills training via publicly-facing websites at accredited medical schools and the academic health sciences libraries serving those medical schools. Phase Two consisted of a survey of the librarians who provide this training to undergraduate medical education students at the Western U.S. medical schools. The survey revealed gaps in forming a complete picture of current practices, thereby generating additional questions that were answered through the Phase Three in-depth interviews.

Results

Publicly-facing websites reviewed in Phase One offered uneven information about library and informatics training at Western U.S. medical schools. The Phase Two survey resulted in a 77% response rate. The survey produced a clearer picture of current practices of library and informatics training. The survey also determined the readiness of medical students to pass certain aspects of the United States Medical Licensure Exam. Most librarians interacted with medical school curricular leaders through either curricula committees or through individual contacts. Librarians averaged three (3) interventions for training within the four-year curricula with greatest emphasis upon the first and third years. Library/informatics training was integrated fully into the respective curricula in almost all cases. Most training involved active learning approaches, specifically within Problem-Based Learning or Evidence-Based Medicine contexts. The Phase Three interviews revealed that librarians are engaged with the medical schools' curricular leaders, they are respected for their knowledge and teaching skills, and that they need to continually adapt to changes in curricula.

Conclusions

This study offers a long overdue, systematic view of current practices of library/informatics training at Western U.S. medical schools. Medical educators, particularly curricular leaders, will find opportunities in this study's results for more productive collaborations with the librarians responsible for library and informatics training at their medical schools. Keywords: Medical libraries, Medical informatics, Teaching, Active learning, Curriculum, Library science, Information science, Information literacy, Information fluency, Information seeking behavior

1. BACKGROUND

Medical students must master skills to retrieve, critically assess, and integrate biomedical information into their clinical decision-making. These skills are recognized as core competencies. As Golub has noted, "The relatively short half-life of medical knowledge has led to the recognition of the importance of instilling the value and the skills of life-long learning as a core piece of medical education" [1]. Accordingly, over the past 75 years academic health sciences librarians have delivered information skills training as part of the formal education of medical students. William Dosité Postell, reporting on a survey conducted during the 1930s, indicated that 50 of the 64 medical schools in the U.S. (78%) offered library instruction [2]. Earl's 1996 report on a survey of 123 academic health sciences libraries produced 55 responses with 75% reporting that they provided library instruction to medical students [3].

The 1982 Matheson Report advised educators that medical education in the future would bear little resemblance to the past due to a daunting expansion of medical information. Future physicians, while still in medical school, would need to acquire a new set of skills to manage and interpret the huge volume of information [4]. The Association of American Medical Colleges' (AAMC) inventory of informatics competencies prompted some academic health sciences libraries in the U.S. to reassess, revamp, and redeploy their library instruction programs to better prepare medical students for a future requiring sophisticated information seeking skills. The arrival of these AAMC competencies generated a great deal of discussion among health sciences librarians, but it remained unclear as to the extent that librarians were ensuring that these AAMC competencies were integrated into medical school curricula [5,6].

Health sciences librarians perform a variety of expected and unexpected roles in U.S. medical school curricula, as validated by an extensive review of studies [7]. Health sciences librarians in the western U.S. have reported on a number of studies that focus on novel or effective library instruction approaches to training medical students at individual academic health sciences libraries [8-32]. No recent surveys have updated Earl's 1996 study, however; and, there is an absence of research that reports comprehensively on the state of library instruction in the western region of the U.S.

Concerns about these research gaps drew the interest of a regional chapter of Libraries in Medical Education (LiME), an interest group of the Association of American Medical Colleges (AAMC) Group on

Educational Affairs. LiME/AAMC meets annually as a means for members to report on current instruction related activities of librarians at institutions in the region. Wishing to take a more systematic and comprehensive approach, in 2009 a LiME research task group undertook an environmental scan of library instruction for medical students at all academic health sciences libraries serving accredited medical schools in the Western United States. The long term goal of the task force was to create a group of interested participants who could support a process of data gathering and reflection on current practices in order to improve the integration of library instruction into medical education. The purpose of this study was to facilitate broad comparisons between peer libraries by exploring in a comprehensive and systematic manner the ways in which academic health sciences libraries in the Western United States deliver instruction to medical students.

2. METHODS

The investigators implemented a three phase action research project consisting of (1) a descriptive environmental scan, (2) survey, and (3) interview methodologies. The present study included the common action research elements of researcher participation, real-life field settings, and reflective periods [33]. Vezzosi's use of an action research approach to understand the effectiveness of library instruction represents a model of how action research can be employed in this subject area [34]. Somekh delineates eight principles normally found in action research in education contexts. The present study incorporated seven of those principles: a cyclical process; collaborative partnerships; knowledge development; roles of the researchers in the process; exploratory engagement; researchers as learners; and a broad contextual awareness [35].

2.1. Phase one

Guided by discussion at LiME meetings and conversations between task force members, Phase One consisted of an unobtrusive environmental scan of publicly facing websites of academic health sciences libraries and educational institutions they serve, focusing on the 17 accredited medical schools of the Association of American Medical Colleges (AAMC) in the Western U.S. listed in Table 1. The investigators sought to construct a detailed picture of educational activities conducted by medical librarians and to identify common patterns of curricular support. Team members made preliminary investigations of public-facing websites at the institutions in the western U.S. Through an iterative process of review, reflection, synthesis, and discussion team members devised a checklist to apply to all 17 sites. This team-generated checklist guided reviewers in examining publicly-available documents such as library newsletters, course guides, and annual reports as well as relevant data from the Association of Academic Health Sciences Libraries (AAHSL) [36]. During the process, the investigators looked for unique or innovative library instruction practices. They also identified basic descriptive information about the user population of the library and, in some cases, information about the faculty status and committee appointments of library staff.

Despite the variable quality and quantity of the initial results, Phase One provided useful information to help investigators articulate the following three research questions to guide phases two and three:

1. What are the current core or commonly followed practices of teaching library/informatics skills to medical students?

2. What patterns or possible trends might emerge from comparisons of different academic health sciences libraries in the Western US that provide library/informatics skills trainings for medical students?

3. What innovative practices can be identified at specific academic health sciences libraries that might be adapted to other academic health sciences libraries?

2. Methods

Table 1. Potential and actual participants in phases 1 & 2: academic libraries supporting schools of medicine

University & Library	Responded to phase 2
1. Charles Drew University of Medicine & Science, Health Sciences Library	✓
2. Loma Linda University Medical Center, Jesse Medical Library & Information Center	✓
3. Oregon Health and Science University, Library	
4. Stanford University Medical Center, Lane Medical Library	✓
5. University of Arizona (Tucson Campus), Arizona Health Sciences Library	
(University of Arizona (Phoenix Campus) Partnership of U of A & ASU medical school dissolved mid-project. ASU counted as part of U of A)	
6. University of California, Davis, Carlson Health Sciences Library	✓
7. University of California, Irvine, Science Library	✓
8. University of California, Los Angeles, Biomedical Library	
(University of California, Riverside program is developing, with most services provided by UCLA and therefore, counted under UCLA)	
9. University of California San Diego, Biomedical Library	✓
10. University of California, San Francisco, Library	✓
11. University of Colorado, Health Sciences Library	✓
12. University of Hawaii at Manoa, Health Sciences Library	
13. University of Nevada Reno, Savitt Medical Library	✓
14. University of New Mexico School of Medicine, Health Sciences Library and Informatics	✓
15. University of Southern California, Norris Medical Library	✓
16. University of Utah, Eccles Health Sciences Library	✓
17. University of Washington, Health Sciences Library	✓

2.2. Phase two

The team shared its analysis of the Phase one data with the larger LiME membership for comment and discussion to guide the design and distribution of a descriptive survey [37]. The survey's final format incorporated the Phase One unobtrusive study data, the investigators' own library instruction experiences, feedback from the (AAMC/LiME) group, and anecdotal knowledge of instructional activities typical in health sciences libraries.

The investigators designed the survey to learn: the medical school governance structure, the role (if any) of librarians in that governance structure, details about library instruction integrated within the curriculum, library instruction (if any) not integrated within the curriculum, faculty status, how library/informatics instruction skills were assessed, and a prediction as to whether graduating medical students at their institution would perform well on PubMed database searches on a United States Medical Licensure Exam (USMLE) currently under consideration by the National Board of Medical Examiners [38,39]. Additional file 1 contains the Phase Two survey questions.

The investigators secured Institutional Review Board approval (HRPO # 10–102) from the University of New Mexico to conduct the survey and any follow-up interviews in the last phase. The investigators deployed the invitation to complete the survey on April 7, 2010. The investigators emailed this invitation to the directors of all 17 academic health sciences libraries serving accredited medical schools in the Western U.S. as listed in the AAHSL Directory [40]. The directors were asked to forward the emailed invitation to all library employees responsible for conducting library instruction with medical students, as a modified form of snowball sampling. A total of three reminder emails were sent and the final invitation was sent in mid-June with an announced closing date of June 22, 2010. The invitation required all respondents to consent to participate in accordance with ethical research principles and invitees were asked to click on a link to the survey as their means of giving consent. Table 1 lists the institutions contacted with checkmarks aside those institutions responding to the survey. The investigators compiled the survey responses, discussed them at length via online conferencing software, and synthesized the data. In keeping with the reflective phase of action research, the results were shared with the librarian community in a panel presentation at a regional meeting of WGEA [41]. The ensuing commentary and discussion among meeting attendees were critical in devising the third phase of the study.

2.3. Phase three

This phase of the project consisted of the investigators developing and deploying a standardized template of six (6) interview questions. The template additionally included some prompts intended to follow these specific questions so the interviewer might pursue any productive avenues for further discussion. The investigators interviewed the respondents at each institution who had the greatest breadth and depth of library instruction experience with medical students. The structured interview questions, and the prompts for possible follow-up, appear in Additional file 2. The investigators implemented the follow-up interviews lasting approximately 30 minutes each by telephone or online conference software beginning in December 2010 and completed the structured interviews during April 2011. All interviewees were sent summaries of the interviews so they might correct any responses, or add clarifying text.

3. RESULTS

This three-phase action research study produced results on the state of library/informatics training that can both inform current practices for medical educators and point toward future research. The environmental scan in Phase One generated targeted research questions about current practices while Phases Two and Three predominantly painted a picture of current practices.

3.1. Phase one results

The information gathered from the websites ranged from ones that merely outlined the essential library services offered extending all the way to websites offering comprehensive accreditation self-study reports in accordance with the standards set by the Liaison Committee on Medical Education guidelines [42]. Inspection of the institutional websites revealed announcements of upcoming workshops, links to handouts from educational sessions and workshops, indications of curriculum-based courses, links to online multimedia tutorials, and access to supplementary instructional guides developed by librarians. While some of the institutions' websites provided a complete picture of their library instruction activities, many lacked sufficient detail to accurately portray the roles that librarians play in supporting medical school curricula. The investigators recognize that some of this information might have been behind password protected websites and thereby unavailable. As noted earlier, the constraints of this purely descriptive approach resulted in an incomplete and inconclusive picture of library instructional programs. An analysis of gaps in the data helped to shape subsequent phases of the study and enabled investigators to generate targeted survey questions intended to yield comparative information about library instruction to medical students.

3.2. Phase two results

Colleagues at 13 of 17 eligible academic health sciences libraries completed the survey, a response rate of 77%. Two librarians from one library completed the survey, and as their responses were consistent with one another, the investigators merged these responses. An informal follow up by one investigator with colleagues at three of the four non-responding libraries revealed that they did not have time to complete the survey. No significant geographic, governance, or other recognizable characteristics distinguished the non-responders from those who responded to the survey. For the responding libraries, all 13 medical schools governed their curricula with a curriculum committee. Academic health sciences librarians interacted with these curriculum committees directly through a variety of methods including regular membership, ex-officio membership, specialized subordinate groups, regular meetings with curricular leaders, or via informal contacts. The plurality of responses indicated that most organizations had multiple means of interaction but the primary method was via ex-officio membership on curriculum committees. A little more than half (53%) of the respondents had faculty status at their respective library and one also had an academic appointment through the school of medicine. All others had academic promotion systems equivalent to faculty status within their institutions. The respondents had an average of 18.4 years of experience as librarians and only three respondents had fewer than 10 years of experience. The majority of respondents had been involved in the most recent Liaison Committee on Medical Education accreditation review process.

The Phase Two survey emphasized identifying instances where librarians engaged in curricular-based library interactions with medical students. All but one of the 13 institutions required incoming medical students to attend basic library orientation sessions. In total, 53 discrete sessions were described along with the year in which the students experienced the sessions. Responses showed activity occurred across the undergraduate curriculum. In general, librarians had an average of three (3) interventions integrated within the core curricula. Not surprising, first year medical students were the target audience for the majority of sessions (29 in total). Third year medical students were the second most frequently contacted audience (21 sessions) followed closely by the second year students (with 16 sessions). See Figure 1. Some sessions were composed of a mix of students from different years. The majority of sessions ($n = 44$) were required with less than 20% ($n = 9$ sessions) as elective sessions]. Fourth year student activity consisted primarily of liaison contacts or consults.

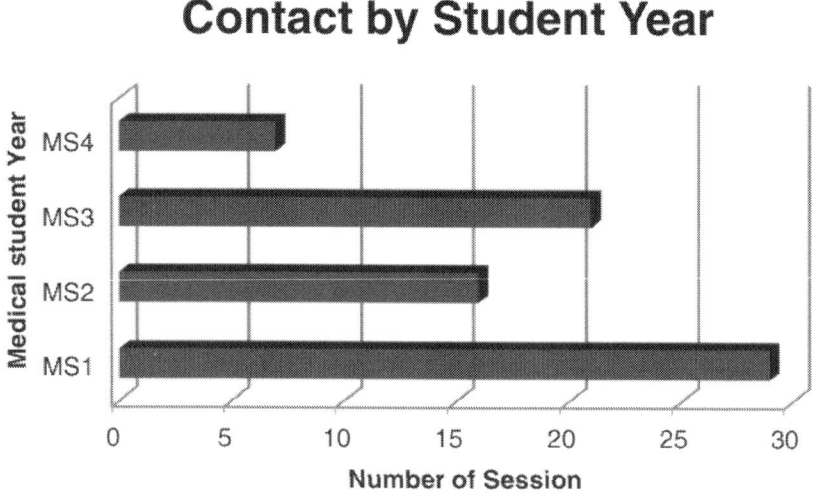

Figure 1. Library instruction by medical student year.

Descriptions of the instruction sessions by respondents were consistent across the different institutions so that, even with institutional variances, responses could be categorized and quantified. Figure 2 summarizes the five (5) types of instruction sessions that emerged: hands-on, lecture, virtual, non-specific orientation, and required consults. Hands-on sessions included anything described with that term or a description indicating student interactions or student practice. Lecture sessions include those described as such as well as ones described as multiple week sessions. Hands-on sessions and lecture sessions were indicated equally with 19 sessions each. Virtual instruction is a growing trend in libraries [43] and the librarian medical educators noted 8 virtual instruction sessions which included work through blogs, online student peer assessment, wikis, videos, or online tutorials. Orientation sessions, not otherwise described, were left as such and termed non-specific orientation. See Table 2.

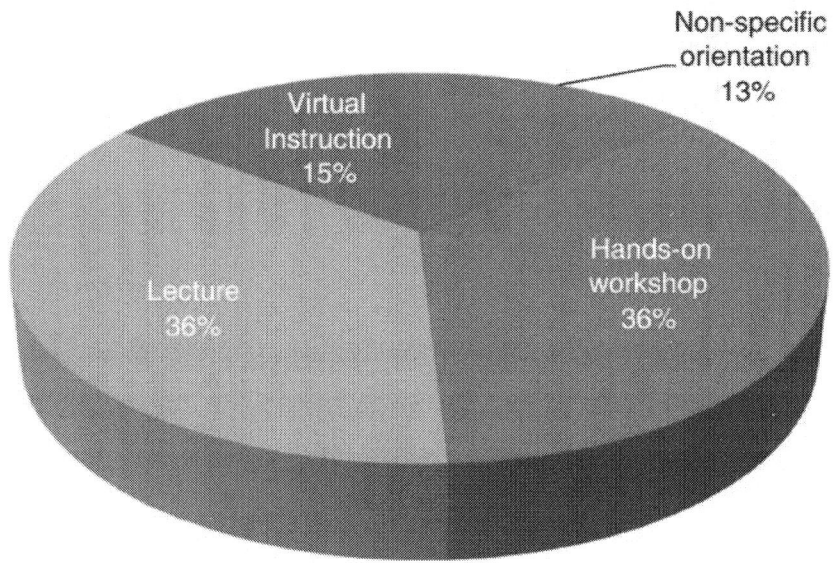

Figure 2. Types of librarian instruction.

Other than ubiquitous PubMed sessions, two distinct topics were volunteered in the descriptions – evidence-based medicine (EBM) (23 sessions) and problem-based learning instruction (5 sessions). Figure 3 indicates that faculty status does not appear to have an impact on curriculum-integrated session *except* that faculty librarians tend to offer a few more required sessions (i.e., fourth and fifth sessions). Only a couple of the non-faculty librarians offered more than three sessions, and these were not always required. One-quarter of the descriptions voluntarily detailed time spent on instruction activities and future iterations of this study might request this specific information. For this small subset, the average time spent on instruction was 2 hours – ranging from a minimum of 30 minutes to a maximum of 32 hours (for a multi-week sequence).

Table 2. Number of instruction sessions by format

Format	Total
Hands-on workshop	19
Lecture	19
Virtual Instruction	8
Non-specific orientation	7
Required Consult	1

Phase two results.

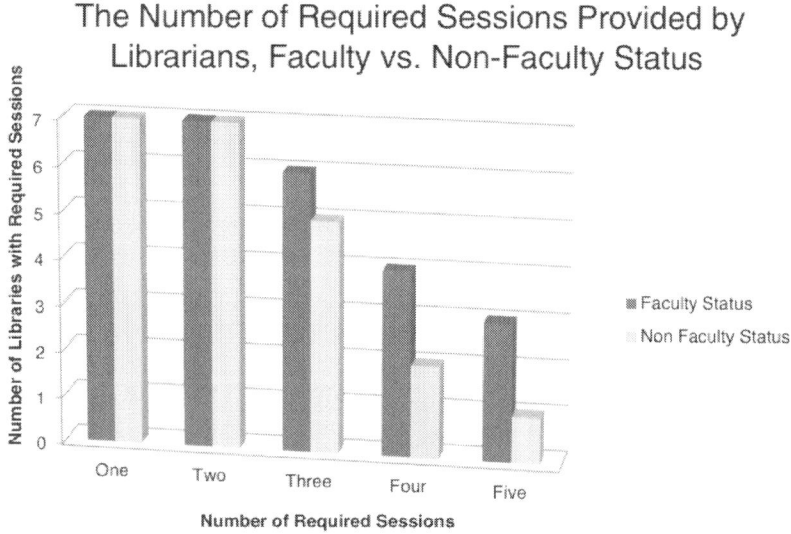

Figure 3. Number of instruction sessions by librarian faculty vs. librarian academics, non-faculty status. Faculty Status, Non-faculty status.

Nearly one-quarter of the libraries reported an assessment of medical students prior to instruction. Formal assessment within the curriculum seems to be a rarely performed activity for librarians, however. As part of the curriculum, schools of medicine have some assessment of knowledge and skills, but it is unclear how the librarians are involved with that activity. Those responding to this question, representing 5 of the 13 schools, submitted just 10 instances of assessment. Yet, as a comparison, over 50 instruction sessions were entered in the survey. Of the assessments, a total of 9 were graded or pass/fail assignments with 2 having a self-assessment or peer-assessment component. Most of the described assessments involved activities such as finding resources and evaluating search skills in order to answer questions. Four sessions dealt specifically with evidence-based medicine (EBM) topics and only two specifically mentioned dealing with citations. The themes identified in Phase Two survey results are consistent with the literature in suggesting that medical students have a diminished preference for non-specific library orientations that lack a curricular context and focused learning objectives.

3.3. Phase three results

During the autumn of 2010, the investigators held several in-depth online conference meetings to discuss the survey results. The structured interview questions to be used in Phase Three emerged from this action research process of review, reflection, and discussion. Twelve (12) of the 13 survey respondents were able to arrange interviews with an investigator during the allotted timeframe, a participation rate of 92%. The 12 interviews occurred during the December 2010 to April 2011 time period. The investigators engaged in both synchronous and asynchronous discussions to reach consensus on their interpretations of these interviews as summarized underneath each of the following italicized questions.

1. *Could you explain the reasons for the successes you have experienced in integrating information literacy/fluency/competencies into your medical school's curriculum?*

Answers varied widely, but some recurring themes emerged from the combined interviews:

- Librarians are engaged with the medical school curriculum committee and with curricular leaders.
- Librarians' efforts frequently rely upon "champions" within the medical school who can advocate for integrating library/informatics skills.
- Librarians have strong support for library or informatics instruction from the library administration.
- Librarians have proven themselves to their teaching faculty colleagues or medical school administrators over time by demonstrating both their knowledge and teaching skills.

2. *If we created a supplement to our upcoming article in a publicly accessible institutional repository that contains samples of outstanding handouts or other documents, would you be willing to contribute 3–5 of your best items?*

- Responses point to a willingness to donate instruction related materials as well as enthusiasm for creating an open access archive.

3. *Could you describe your online curricular or instructional support (examples: learning management system such as Blackboard; social networking; chat) at your institution? Does the library or another unit such as IT provide this support?*

Most medical schools use a commercial learning management system. Most also use a locally produced learning management system to supplement the commercial system in order to meet all of their needs.

4. *What were the "lessons learned" from past mistakes or miscalculations in your efforts?*

- Free-standing courses never work as well as library instruction that is integrated fully into the curriculum
- The need to keep adapting to changing circumstances, including curricular changes, in the medical school
- Secure detailed feedback from students on the quality of teaching, its relevance to curricular content, and the content taught
- Perseverance despite setbacks usually leads to success

5. *Why are librarians at your library motivated to teach?*

Most librarians taught because of their faculty status, or were expected to teach due to a similar codified equivalent of faculty status as an institutional career ladder for promotion. Beyond this broad expectation, however, respondents noted that most librarians teach as a natural outgrowth of their desire to ensure that medical students (and future physicians) possess all needed library/informatics skills. One respondent mentioned that there were too few librarians to teach these skills on an individual point-of-use basis so formal instruction was the only reasonable cost-effective option. Interestingly, multiple authors made this central cost-effectiveness argument in a classic volume published in 1974 during a renaissance within library instruction in academic libraries [44]. Additionally, most respondents indicated that those librarians who teach certainly enjoy this instructional role.

6. *Reviewing your responses concerning your activities, how much time was devoted to each?*

Respondents' formal work allocation to the education of medical students encompassed anywhere on average from 15 to 50% of their overall efforts. Most respondents reported that they spend a considerable amount of time outside of the classroom with curricular design, keeping abreast of curricular changes, and preparing to teach. On this latter point, one respondent mentioned spending 25 hours to perfect a presentation for a single one hour session in front of medical students since she realized that her time was so limited within today's "crowded curriculum" [45] at US medical schools.

Figure 4 provides a Wordle™ word cloud that visually displays the words used most frequently by interview respondents. The investigators expected to find words such as "librarians" and "library" prominently displayed in the word cloud. The investigators did not expect to see the words "teaching", "medical students", "curriculum", or "faculty" so frequently mentioned. Thus, the word cloud discovered some less obvious patterns otherwise lost by reading the texts of the structured interviews compiled in Phase Three.

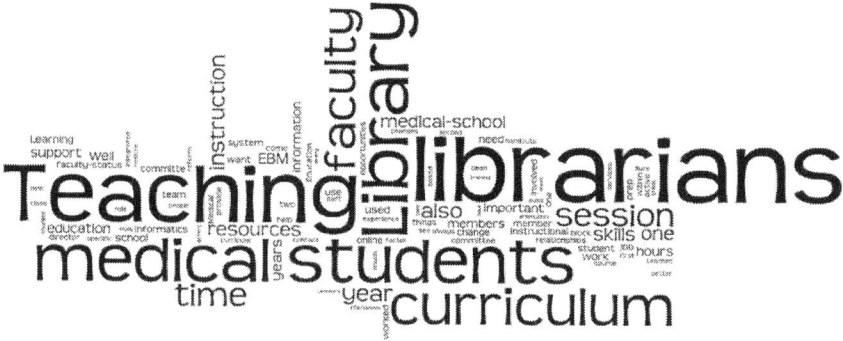

Figure 4. Wordle cloud from phase 3 interviews.

4. DISCUSSION

This study fills a gap in health sciences library/informatics skills training at different US medical schools. The investigators discovered that many of their colleagues achieved success in integrating library and informatics skills into their respective curricula. This project readily confirmed the

4. Discussion

diversity of practices. This study also produced suggestive non-statistical evidence for librarians' status and roles in curricular governance.

4.1. An action research approach

Consistent with the tenets of an action research approach, the investigators followed an evolutionary, developmental course in order to better explore the challenges facing library/informatics instructors in medical education. By examining and sharing data in a stepwise approach, investigators were able to integrate discussion and concerns from the practitioner community in order to improve each subsequent phase of inquiry. This iterative process of engagement contributed significantly to the intended goal of producing a useful report on practices and trends in library instruction.

Because the investigators were members of the very AAMC/LiME community of practitioners under study, they bridged conventional forms of dichotomy between themselves and their subjects, contributing to a collaborative co-construction of knowledge [33]. Incorporation of key aspects of action research in the study, including building relationships, acknowledging and sharing power, and encouraging participation of the study population [34], enhanced the eventual applicability of results to professional practice.

4.2. Phase one

This phase revealed that an institution's publicly facing website cannot be relied upon to gather enough data to make more than just superficial comparisons across institutions on library education. The investigators learned in this process moreover that the availability of more robust data would be inconsistent across institutions, at best.

4.3. Phase two

The survey addressed many of the investigators' questions generated during Phase One. A particular focus examined the extent to which curriculum integration is reflected in library instructional activities. The literature has long suggested that increased educational effectiveness and impact on student learning is predicated on integration of library instruction into the existing medical curriculum, rather than as a separate component of a library's educational program [46].

In her landmark article, Francesca Allegri defines "course integrated instruction" as having met at least three of the four following criteria: "(1) faculty outside the library are involved in the design, execution, and evaluation of the program, (2) the instruction is curriculum-based, in other words, directly related to the students' course work and/or assignments, (3) students are required to participate, and (4) the students' work is graded or credit is received for participation [47]." Survey responses reflected and met Allegri's definition of course integrated instruction. Respondents described a variety of curriculum integration activities such as recurring roles in semester long classes; collaborative teaching of informatics concepts to support problem-based learning exercises, and interactive instruction covering content tied directly to exam questions. EBM training has evolved over the past few decades with librarians having a growing role in working with both students and faculty within the curriculum [48,49]. The survey responses indicated that over 40% of the sessions were EBM topics, a finding that validates much of the research in this area.

Faculty status or its close equivalent for librarians appears to provide access and credibility for librarians needing to integrate library or informatics training into medical school curricula. Librarians and teaching faculty members alike seemed to recognize their mutual interdependence in these endeavors. One of the founders of the modern library instruction movement, Evan Farber, has emphasized this mutually-dependent relationship between librarians and their teaching faculty colleagues [50]. Travis has admonished her colleagues more recently that "Librarians need to think and act globally, never compartmentalize library instruction efforts, and find ways to scale information literacy into an institution wide model [51]." Librarians at the institutions in this study apparently were paralleling Travis' advice as further evidenced by their successes. Librarians involved in providing integrated library and informatics instruction had an average of 18 years' experience, which strongly suggests that this role requires considerable experience, knowledge, and expertise. Wiggins similarly has noted that library or informatics instruction often succeeds when the experienced and knowledgeable librarian can provide skeptical students with the rationales for the relevance of library instruction at a specific juncture in the curriculum [52]. The recent resurgence of interest among educators on the national level in cultivating affective educational objectives also dovetails with this data [53].

4.4. Phase three

The Phase Three interviews highlighted the importance of having champions among the teaching faculty and the support of administrators overseeing the curriculum. Curzon has emphasized the importance of such partnerships, particularly with teaching faculty who must balance a crowded curriculum with the student's escalating need to effectively manage the exploding information universe [54]. In the absence of a context or perceived need among the students, interview respondents reported that the basic library orientation sessions tend to have poor educational outcomes. Prior research had suggested that library/informatics instruction most likely will be more effective when integrated into the curriculum [55]. This study preceded publication of Moore's 2011 sentinel *Academic Medicine* commentary on the need for library/informatics training. The findings in this study provide supplementary evidence to support Moore's thesis [56] as well as revealed innovative ways librarians are maximizing limited instruction time with their curricular partners.

4.5. Limitations

This study details an environmental scan that explored the breadth and depth of library/informatics skills instruction for medical students at academic health sciences libraries in the Western U.S., and represents a unique examination of a largely uncharted subject area. The authors could identify only one account that bore even a distant similarity to the approach found in the present study [57]. The research reported in this present study cannot be generalized to the entire U.S. due to the geographic concentration in the western region, the small number of institutions, and the investigators' awareness of diverse library instruction practices in other regions. The survey responses also constituted low-level frequency and descriptive data that could not be easily categorized into discrete data points. Still, medical educators and librarians outside the region can benefit from learning about the rich and diverse descriptive information on how their colleagues at different western U.S. institutions grapple with challenges similar to their own. In the process of implementing this three phase action research project the investigators have created a template for a national level action research study. This template could even be modified to secure more defined responses, if viewed by colleagues elsewhere as desirable. The investigators would be happy to share with interested colleagues their experiences in conducting this type of multiple methods study.

4.6. Future research

Expanding the focus of this research beyond the Western region would provide a sufficient sample of librarians to make statistically significant test of the following hypotheses:

1. Great diversity in how medical students are trained on library/informatics skills exists in the United States, and that knowledge of some of these practices will be valued by colleagues involved in similar types of library/informatics training.

2. A correlation exists between librarian roles in governance structures *and* their degree of involvement in training medical students on library/informatics skills, the degree to which this training has been integrated into the curriculum, and their assessment of medical student performance.

5. CONCLUSION

This study provides medical educators and librarians with a detailed snapshot illustrating the current nature of library instruction in medical schools. It delineates the degree to which these library/informatics competencies are integrated into medical school curricula. Analysis of the information examines some preconditions for successful instructional programs, reveals challenges shared by librarian instructors, and discusses adaptive strategies that have led to greater student satisfaction. The results reinforce the notion that information skills instruction is an important part of medical education and are indicative of the value librarians contribute to the educational process.

Medical educators, if not already doing so, should actively partner with librarians at their institution to strive for curriculum integrated information skills training of medical students. Librarians should also ensure that feedback on library instruction is included as a standard component of student course evaluations. Folding evaluations of library instruction into the broader curricular context may increase the validity of student feedback, give instructors meaningful data with which to quantify skills improvement, enhance future library instruction, and relieve students of the burden of completing separate post-instruction library surveys. Librarians play a pivotal role in providing the skills to bolster life-long learning that goes well beyond medical school and prepares a solid foundation for how to keep up with the ever-growing body of medical education research literature.

AUTHORS' CONTRIBUTIONS

JE, KH, TH, and JT conceived of this project. JE developed the design and secured IRB approval and subsequent renewals. JE, KH, TH, and JT conducted the Phase One environmental scan and designed the Phase Two survey. KH implemented the survey via SurveyMonkey™ and codified and tabulated the results. JE, KH, and TH interpreted the survey results and designed the follow-up interview questions in Phase Three. JE, KH, TH, and JT conducted the in-depth interviews in Phase Three. JE, KH, TH, and JT wrote and edited the manuscript. All authors read and approved the final manuscript.

ACKNOWLEDGEMENTS

The authors wish to thank the librarians at the AAMC's LiME WGEA 2009 meeting for the inspiration for this project. Thanks also to all the librarians who provided such rich information on the survey and interviews.

REFERENCES

1. Golub RM. Medical education theme issue 2008: call for papers. JAMA. 2007;298(22):2677.

2. Postell WD. Further notes on the instruction of medical school students in medical bibliography. Bull Med Libr Assoc. 1944;32(2):217–220. [PMC free article] [PubMed]

3. Earl MF. Library instruction in the medical school curriculum: a survey of medical college libraries.Bull Med Libr Assoc. 1996;84(2):191–195. [PMC free article] [PubMed]

4. Matheson NW, Cooper JA. Academic information in the academic health sciences center. roles for the library in information management. J Med Educ. 1982;57(10 Pt 2):1–93.

5. Contemporary issues in medical education: medical informatics and population health.https://members.aamc. org/eweb/upload/Contemporary%20Issues%20in%20Med%20Medical%20Informatics%20ReportII.pdf.

6. McGowan JJ, Passiment M, Hoffman HM. Educating medical students as competent users of health information technologies: the MSOP data. Stud Health Technol Inform. 2007;129(2):1414.

7. Schwartz DG, Blobaum PM, Shipman JP, Markwell LG, Marshall JG. The health sciences librarian in medical education: a vital pathways project task force. J Med Libr Assoc. 2009;97(4):280–284.

8. Kingsley K, Galbraith GM, Herring M, Stowers E, Stewart T, Kingsley KV. Why not just google it? An assessment of information literacy skills in a biomedical science curriculum. BMC Med Educ.2011;11:17.

9. Geppert CM, Arndell CL, Clithero A, Dow-Velarde LA, Eldredge JP, Kalishman S, Kaufman A, McGrew MC, Snyder TM, Solan BG. et al.

Reuniting public health and medicine: the university of new mexico school of medicine public health certificate. Am J Prev Med. 2011;41(4 Suppl 3):S214–S219.

10. Dodson S, Gleason AW. Web 2.0 support for residents' and fellows' patient care and educational needs. Med Ref Serv Q. 2011;30(2):95–101.

11. Pozdol JR. Ten steps to increase library impact on an academic health sciences campus. Med Ref Serv Q. 2010;29(3):229–239.

12. Kroth PJ, Phillips HE, Eldredge JD. Leveraging change to integrate library and informatics competencies into a new CTSC curriculum: a program evaluation. Med Ref Serv Q. 2009;28(3):221–234.

13. Jeffery KM, Maggio L, Blanchard M. Making generic tutorials content specific: recycling evidence-based practice (EBP) tutorials for two disciplines. Med Ref Serv Q. 2009;28(1):1–9.

14. Eldredge JD. Student Peer Assessment as an Instructional Strategy. Albuquerque, NM: LOEX 37th National Conference; 2009.

15. Chen HC, Tan JP, O'Sullivan P, Boscardin C, Li A, Muller J. Impact of an information retrieval and management curriculum on medical student citations. Acad Med. 2009;84(10 Suppl):S38–S41.

16. Eldredge JD, Carr R, Broudy D, Voorhees RE. The effect of training on question formulation among public health practitioners: results from a randomized controlled trial. J Med Libr Assoc.2008; 96(4):299–309.

17. Ryce A, Dodson S. A partnership in teaching evidence-based medicine to interns at the university of Washington medical center. J Med Libr Assoc. 2007;95(3):283–286.

18. Teal J, Eldredge J. Staff Development. Higginbottom PC. Chicago: Medical Library Association; 2005. The University of New Mexico.

19. Reavie K, Persily GL, Souza KH. In: A guide to developing end user education programs in medical libraries. Connor E, editor. New York: Haworth Information Press and Haworth Medical Press; 2005. Integrating Medical Informatics into the School of Medicine Curriculum at the University of California, San Francisco; pp. 209–225.

20. Owen DJ, Persily GL, Babbitt PC. In: A guide to developing end user education programs in medical libraries. Connor E, editor. New York: Haworth Information Press and Haworth Medical Press; 2005. An Informatics Course for First-Year Pharmacy Students at the University of California, San Francisco; pp. 129–143.

21. Eldredge JD. In: Informatics in health sciences curricula. Sewell RR, Brown JF, Hannigan GG, editor. Chicago: Medical Library Assn; 2005. EBM informatics component of the Genetics & Neoplasia Block.

22. Eldredge JD. Search strategies for population and social subjects in a medical school curriculum. Med Ref Serv Q. 2004;23(4):35–47.

23. Eldredge JD. The librarian as tutor/facilitator in a problem-based learning (PBL) curriculum. Ref Serv Rev. 2004;32(1):54–59.

24. Brown JF, Nelson JL. Integration of information literacy into a revised medical school curriculum.Med Ref Serv Q. 2003;22(3):63–74.

25. Kaplowitz J, Wilkerson L. Reaching and teaching new medical students. Acad Med.2002;77(11):1173.

26. Kaplowitz JR, Yamamoto DO. Web-based library instruction for a changing medical school curriculum. Libr Trends. 2001;50(1):47–57.

27. Eldredge JD, Rhyne RL. In: Handbook on problem-based learning. Rankin JA, editor. Chicago: Medical Library Assn; 1999. Building foundations for effective library skills in medical education:

library/biometry projects in the first month of medical school; pp. 407–432.

28. Eldredge JD, Teal JB, Ducharme JC, Harris RM, Croghan L, Perea JA. The roles of library liaisons in a problem-based learning (PBL) medical school curriculum: a case study from university of New Mexico. Health Libr Rev. 1998;15(3):185–194.

29. Owen DJ. Using personal reprint management software to teach information management skills for the electronic library. Med Ref Serv Q. 1997;16(4):29–41.

30. Butros A. Using electronic mail to teach MELVYL MEDLINE. Med Ref Serv Q. 1997;16:69–75.

31. Minchow RL, Pudlock K, Lucas B. Breaking new ground in curriculum integrated instruction. Med Ref Serv Q. 1993;12:1–18.

32. Eldredge J. A problem-based learning curriculum in transition: the emerging role of the library. Bull Med Libr Assoc. 1993;81(3):310–315.

33. Hannigan GG. Action research: methods that make sense. Med Ref Serv Q. 1997;16(1):53–58.

34. Vezzosi M. In: Chandos information professional series. Connor E, editor. Oxford: Chandos Pub; 2007. Evidence-based librarianship : case studies and active learning exercises; pp. 19–40.

35. Somekh B. In: The Sage handbook of educational action research. Noffke SE, Somekh B, editor. Thousand Oaks, CA: Sage Publications Ltd; 2009. Introduction; pp. 1–9.

36. Association of Academic Health Sciences Libraries. Annual Statistics of Medical School Libraries in the United States and Canada. 32. Seattle: AAHSL; 2010.

References

37. Eldredge JD, Heskett KM, Henner T, Tan J. New Horizons for Integrating Library/Informatics Skills into Medical Curricula: Report of the 2010 LiME Research Study. Stanford, CA: New Horizons: Selecting, Teaching and Inspiring the Next Generation of Physicians Association of American Medical Colleges Western Group on Educational Affairs (AAMC/WGEA) Regional Conference; 2011.

38. National Board of Medical Examiners. USMLE moves to next step in design, review. Examiner.2008;55(2):1–4.

39. Anderson OS. Changing the USMLE: challenges and opportunities for physiology and other medical school basic science departments. Physiologist. 2009;52(2):39–44.

40. Association of Academic Health Sciences Libraries. AAHSL Membership Directory 2009 and AAHSL 30th Annual Report, 2007–2008. Seattle, WA: Association of Academic Health Sciences Libraries; 2009. pp. 5–29.

41. Eldredge JD, Henner T, Heskett K, Tan J. Current Practices in Library/Informatics Instruction in Academic Libraries Serving Medical Schools in the Western US. Asilomar, CA: Health and Interprofessional Education for the Underserved: Model Programs and Innovations Association of American Medical Colleges Western Group on Educational Affairs (AAMC/WGEA) Regional Conference; 2010.

42. Liaison Committee on Medical Education. Functions and Structure of a Medical School. Standards for Accreditation of Medical Education Programs Leading to the M.D. Degree. 2010.http://web.archive.org/web/20110522235937/http://www.lcme.org/functions2010jun.pdf.

43. Gardois P, Colombi N, Grillo G, Villanacci MC. Implementation of Web 2.0 services in academic, medical and research libraries: a scoping review. Health Info Libr J. 2012;29(2):90–109.

44. Lubans J. Educating the library user. New York: R. R. Bowker Co; 1974.

45. Smith HC. A course director's perspectives on problem-based learning curricula in biochemistry. Acad Med. 2002;77(12 Pt 1):1189–1198.

46. Burrows S, Ginn DS, Love N, Williams TL. A strategy for curriculum integration of information skills instruction. Bull Med Libr Assoc. 1989;77(3):245–251.

47. Allegri F. Course integrated instruction: metamorphosis for the twenty-first century. Med Ref Serv Q.1986;4(4):47–66.

48. Traditi LK, Le Ber JM, Beattie M, Meadows SE. From both sides now: librarians' experiences at the rocky mountain evidence-based health care workshop. J Med Libr Assoc. 2004;92(1):72–77.

49. Dorsch JL, Jacobson S, Scherrer CS. Teaching EBM teachers. Med Ref Serv Q. 2003;22(2):107–114.

50. Farber E. Faculty-librarian cooperation: a personal retrospective. Ref Serv Rev. 1999;27(3):229–234.

51. Travis TA. Librarians as agents of change: working with curriculum committees using change agency theory. New Dir Teach Learn. 2008;2008(114):17–33.

52. Wiggins ME. Instructional design and student learning. Ref Serv Rev. 1999;27(3):225–228.

53. Krathwohl DR, Bloom BS, Masia BB. Taxonomy of educational objectives: the classification of educational goals. Handbook 2: Affective domain. New York: McKay; 1964.

54. Curzon SC. In: Integrating information literacy into the higher education curriculum: practical models for transformation. 1.

Rockman IF, editor. San Francisco: Jossey-Bass; 2004. Developing faculty-librarian partnerships in information literacy; pp. 29–45.

55. Shurtz S. Thinking outside the classroom: providing student-centered informatics instruction to first- and second-year medical students. Med Ref Serv Q. 2009;28(3):275–281.

56. Moore M. Teaching physicians to make informed decisions in the face of uncertainty: librarians and informaticians on the health care team. Acad Med. 2011;86(11):1345.

57. Lodenius L, Honkanen M. Medical information specialist as teacher: teaching searching skills. J Eur Assoc Health Inf Libr. 2011;7(3):3–12.

CHAPTER 6

Clinical and academic use of electronic and print books: the Health Sciences Library System e-book study at the University of Pittsburgh

Barbara L. Folb[1], MM, MLS, MPH, Charles B. Wessel, MLS; Leslie J. Czechowski[2], MA, MLS

[1]*NLM Informationist Fellow, Graduate School of Public Health, Department of Behavioral and Community Health Sciences, University of Pittsburgh, A716 Crabtree Hall, 130 DeSoto Street, Pittsburgh, PA, 15261*
[2]*Assistant Director of Collections and Technical Services; Health Sciences Library System, University of Pittsburgh, 200 Scaife Hall, 3550 Terrace Street, Pittsburgh, PA 15261*

ABSTRACT

Objectives

The purpose of the Health Sciences Library System (HSLS) electronic book (e-book) study was to assess use, and factors affecting use, of e-books by all patron groups of an academic health sciences library serving both university and health system–affiliated patrons.

Methods

A web-based survey was distributed to a random sample (n=5,292) of holders of library remote access passwords. A total of 871 completed and 108 partially completed surveys were received, for an approximate response rate of 16.5%–18.5%, with all user groups represented. Descriptive and chi-square analysis was done using SPSS 17.

Results

Library e-books were used by 55.4% of respondents. Use by role varied: 21.3% of faculty reported having assigned all or part of an e-book for class readings, while 86% of interns, residents, and fellows reported using an e-book to support clinical care. Respondents preferred print for textbooks and manuals and electronic format for research protocols, pharmaceutical, and reference books, but indicated high flexibility about format choice. They rated printing and saving e-book content as more important than annotation, highlighting, and bookmarking features.

Conclusions

Respondents' willingness to use alternate formats, if convenient, suggests that libraries can selectively reduce title duplication between print and e-books and still support library user information needs, especially if publishers provide features that users want. Marketing and user education may increase use of e-book collections.

1. INTRODUCTION

For more than a decade, health sciences libraries have been building and providing digital electronic collections of journals and books [1, 2]. Most print journals have been replaced with electronic journals (e-journals) and readily embraced by users [3, 4]. Academic health sciences libraries continue to expand electronic book (e-book) availability and invest in improving e-book access [5–7], as external book circulation continues to decrease [8]. Book publishers and users are adapting to the e-book format as librarians attempt to determine which book formats, print or electronic, make the most sense for their collections, budgets, and most importantly, their users [9–12].

The scenario, described above, was the impetus for a study by the Health Sciences Library System (HSLS) at the University of Pittsburgh (Pitt) of how geographically distributed and diverse patrons use e-books. HSLS wanted to know if duplication of titles in print and e-book format could be reduced, while still meeting users' information needs. HSLS serves Pitt's six schools of the health sciences <http://www.health.pitt.edu> (medicine, dental medicine, nursing, pharmacy, public health, and rehabilitation) and the hospitals and programs of the UPMC health system <http://www.upmc.com>. HSLS provides access to licensed electronic resources for all of UPMC in the United States and abroad. At the time of the study, HSLS consisted of three libraries: Falk Library, serving the six schools of the health sciences and the UPMC hospitals contiguous to Pitt's main campus; the professional and consumer health libraries at UPMC Shadyside <http://www.upmc.com/HospitalsFacilities/Hospitals/Shadyside/>; and the libraries of Children's Hospital of Pittsburgh of UPMC <http://www.chp.edu>.

HSLS has been collecting e-books for over 10 years and at the time of this study licensed over 2,000 e-books from vendors such as Ovid, MD Consult, STAT!Ref, McGraw-Hill, and Rittenhouse. All HSLS e-books had MARC records in the online library catalog and were included in a browsable list on the libraries' website. A federated clustering search tool developed by the HSLS provided direct access to many of the e-books from the libraries' home page <http://www.hsls.pitt.edu> [6]. Additionally, university faculty, staff, and students could access Pitt University Library System e-books from providers such as ebrary, netLibrary, Springer, and Knovel. At the time of the study, handheld e-book readers such as the Amazon Kindle were rising in popularity for leisure reading purposes. Because academic application of these readers was in its infancy, they were not supported by HSLS [13–15].

2. LITERATURE REVIEW

2.1. Focus of existing e-book literature

Articles on e-books in academic libraries began appearing soon after vendors began offering them in the late 1990s [16]. The literature on e-books in libraries has covered a handful of core issues from the beginning, but the discussion has changed as e-book features, cataloging practice, and user awareness and adoption of e-books has evolved. Issues addressed include identification of e-book users by demographic groups, cataloging practice, and e-book access provision [6, 17–21]; meaningful comparison of statistics on use of print and e-books [9, 22–24]; variability in statistics provided by e-book vendors [25]; activities supported by e-book use [26–30]; use by type of book [31]; user characteristics affecting e-book use [28, 32–34]; and features desired in e-books [5, 16, 27, 30–32, 35, 36].

E-book surveys of academic populations in the United States and the United Kingdom have included multi-university surveys of people in all roles [32] and surveys of all roles at a single university [18, 30, 31, 37], students at a single university [38], students at multiple universities [35], and faculty at multiple universities [27]. E-book studies in academic health sciences environments included a statistical comparison of print book circulation to e-book access [9], a study of how many titles from the Brandon/Hill list were available as e-books in 2004 [7], a survey of medical students in clinical rotations [26], surveys of dental and nursing students using digital textbooks [29, 39], focus groups with midwifery students in the United Kingdom [40], and an observation and interview study including five undergraduate nursing students in the population [41]. No surveys were found that studied all user groups in an academic health sciences library.

2.2. Factors associated with differences in e-book use

When comparing use of print and e-book versions of the same title, studies indicate that e-books are accessed more frequently than print books are checked out [9], but it is difficult to make meaningful comparisons because usage statistics measure different types of access [9, 22–24]. Differences in e-book use by academic discipline [33] or role [30] have been noted but not explored in depth. Studies have not reported significant differences in use between faculty and students. For example, Levine-Clark reported use between 51% and 54% for both groups [33]. There are some modest differences in e-book use between men and women. A UK study found men (65.4%) more likely than women (63.6%) to use e-books, and men were also more likely than women to read a whole chapter on screen [32]. No studies of e-books in hospitals were located that surveyed and compared all hospital employee groups.

2.3. Barriers and facilitators to e-book use

Low awareness of the collections could be a barrier to use. Existing studies reported that 43%–67% of library patrons were aware of library e-book collections [30, 33, 35, 38]. Use of e-books was slightly lower than awareness, varying between 40% and 62% [32, 33, 35, 38, 42].

High e-book visibility and ease of access should increase use. Discovery of e-books occurs through library websites, catalogs, and library staff

[5, 27, 32, 33, 35, 42]. Adding MARC records to the catalog [5, 17, 30] and e-book federated search tools on a website [30] increases use of library e-books. A study assessing e-book accessibility on Association of Research Libraries websites reported that library catalogs usually provided cumbersome, multistep methods to limit a search to e-books. The researchers concluded that catalogs should be modified to provide a single-step e-book limit, and alternative access points to e-books on library websites were needed [19].

Librarians may purchase e-books with distant users in mind, and a study of e-book use by on and off campus students did report that off campus use was disproportionately high [28]. However, a study at Texas A&M University reported that use of e-books by distance students was much lower than their use of e-journals and databases [43]. E-books, just like e-journals, are used by people on campus who can visit the library. One study showed that most researchers used e-books from nonlibrary locations on campus [34]. The convenience of using e-books is appreciated by users both on and off campus.

One frequently discussed barrier to e-book use is the discomfort of screen reading. Because academic users commonly use e-books for ready reference, screen reading may be less of a barrier to adoption than it initially appears. Users prefer to read short sections of books online but prefer print for reading an entire book [30,32]. A UK study reported that 62.6% of students and 57.8% of faculty read entirely on screen the last time they used an e-book; only 6.4% and 6.5%, respectively, printed materials to read [32].

Users expect e-books to include the features and functionality that they enjoy in print, enhanced with online features. Features and attributes desired by academic users included keyword searching, 24/7 accessibility, simultaneous users, downloading, copying and pasting, and printing [30, 42, 44]. Products such as SpringerLink—which have "journalized" e-books, allowing printing and saving of entire chapters and inexpensive print on demand of entire e-books—may make e-books a more well-rounded product [5]. In the 2007 ebrary faculty survey, over half of respondents said the ability to download and fewer restrictions on printing and copying would make e-books more suitable for use [27]. Highlighting and annotation of e-books was desired by 94.9% of students in one survey [36], but ranked lower than searching, access, downloading, and printing features to students in another survey [35].

2.4. Type of book and format preference

No matter if a book is published as a textbook, reference source, or other book type, most readers report using e-books like reference books. They search for specific information and read short sections of needed information across all types of books [32, 33, 39]. Some studies reported e-book use by type. Faculty and students at one UK university reported using e-books as textbooks (59.9%), reference books (52.4%), and research monographs (46%) [42]. No studies were found reporting use of e-books in health sciences settings by type of book.

2.5. Intended use and e-book or print preference

In academia, library e-books are used more for research and individual study than assigned class readings [30]. The ebrary faculty survey found more faculty assigned e-journal readings (57%) than assigned e-book readings (29%) [27]. Research looking at e-textbooks as replacements for student purchase of required class texts reported that integrating e-books into the virtual learning environment was challenging [29], and student success in using them depended on good user education [45].

In the clinical environment, the University of Iowa compared medical student use of print medical books and 3 online resources (UpToDate, MD Consult, and Harrison's Online) to support patient care and learning during clinical rotations. Engaged in intensive learning, students often consulted major medical publications daily in their preferred format. They preferred UpToDate (53%) and MD Consult (33%) by a wide margin over print (14%) [26]. The authors concluded that accessibility was not the main factor driving their preference, because both print and e-books were accessible in the hospital wards. Their survey suggested that student choices are driven by their perception of how quickly they can find an answer.

After reviewing the literature, it was not clear how to apply the findings of these studies to the HSLS user population, as most previous studies did not address health sciences library users. HSLS also had additional questions concerning the use made of e-books and print books in the collection and what features of e-books users specifically wanted and preferred. To work toward an ideal collection that would meet the needs of all users, HSLS developed a survey of its user population.

3. METHODS

3.1. Survey development for the Health Sciences Library System (HSLS) e-book study

A probability sample survey was developed for online administration. Some questions applicable to any library setting were adapted from existing e-book surveys, [32, 33, 35], while other questions specific to health sciences libraries were created for this survey.

The purpose of the HSLS e-book study was to ascertain: (1) what factors and demographic profiles were associated with the differences in print and e-book use, (2) what barriers and facilitators to e-book use did HSLS patrons experience, (3) whether there was an association between the type of book (i.e., manual, textbook, handbook) and format preference (electronic or print), and (4) whether there was an association between

3. METHODS

selecting a print versus an e-book and the patron's intended use (i.e., classroom teaching, clinical, study, research)? Within these four main questions, HSLS identified important sub-questions to address in the survey:

1. What factors and demographic profiles are associated with differences in print and e-book use?

- Are organizational affiliations, roles, or other demographic factors associated with variations in e-book use?

2. What barriers and facilitators to e-book use do HSLS patrons experience?

- Are HSLS users aware of the e-book collection?
- What are user opinions of the e-book discovery and searching tools?
- Does use of the physical library and HSLS website vary with distance from a library or perception of available time?
- How much do HSLS users value e-book features, including full-text searching, saving and printing options, and highlighting and annotation?

3. Is there an association between the type of book and e-book or print preference?

- If replacing print books with e-books, what types of print books would be more acceptable as e-books?
- How flexible are HSLS users about choice of print versus e-book?

4. Is there an association between the selection of a print or e-book and the patron's intended use?

- Is the e-book collection supporting all academic and clinical tasks?

Two versions of the survey (Appendix, online only), one for UPMC with 46 questions and one for University of Pittsburgh with 47 questions, were created to reduce the burden on respondents of non-applicable questions. To reduce confusion, e-book was defined in the survey introduction, and examples of each type of book (reference, textbook, etc.) were given in the appropriate questions. Questions that identified respondents with multiple affiliations and roles, such as a UPMC clinical physician with a faculty appointment at the university were included in both versions. Surveys were entered into Opinio survey software, version 4.3.4; tested by HSLS librarians and graduate students in a survey methods class; and edited based on their feedback. The study received University of Pittsburgh Internal Review Board approval as an exempt study.

3.2. Sampling and survey distribution

The target population was all HSLS library users. This community encompasses faculty, researchers, clinicians, residents, fellows, employees, and students practicing and learning in the schools of the health sciences and across UPMC. The email addresses of all patrons registered for an HSLS remote access password as of March 5, 2009, with librarians removed, served as the sampling frame. In total, there were 5,222 UPMC and 4,250 university email addresses. Sample sizes were calculated assuming 50% of respondents used e-books, adjusted for population size, and an estimated 25% response rate [46], giving samples of 2,608 for the university and 2,684 for UPMC. Random samples were drawn from each list using SPSS 17.0. Email invitations were sent out in March 2009, followed by 3 reminders at 5-day intervals. Data collection continued for 22 days.

3.3. Analysis

The data from both surveys were exported from Opinio into SPSS 17.0 and merged into one file for analysis. Open-ended responses were analyzed and recoded into discrete categories. For example, all nursing specialties respondents who chose to enter under "Other" were recoded as "Nurse." Some response categories, such as the Pitt roles "Staff" and "Research assistant," were combined into one category to yield logical categories of sufficient size for statistical analysis. Basic descriptive statistics and cross-tabulations were run.

4. RESULTS

4.1. Response rate

Response rates were: Pitt, 434 complete, 42 partial responses, total university response, 476; UPMC, 437 complete, 66 partial responses, total UPMC response, 503; and overall response, 871 complete, 108 partial responses, for a total of 979. Missing data from partial responses were handled with pairwise deletion, allowing the use of data from incomplete surveys in the analysis. American Association for Public Opinion Research (AAPOR) methods of response rate (RR) calculation RR5 and RR6 were used to calculate response rates [47], which can only be approximated for this type of email survey [48]. The combined response rate estimate is between 16.5% and 18.5% (university, 16.6%–18.3%; UPMC, 16.3%–18.7%). Response rates to Internet surveys vary widely, as shown by a meta-analysis of Internet survey response rates that reported rates between 7% and 88% with a mean of 34% [49].

4.2. Profile of respondents

Table 1 (online only) presents demographic data on the respondents. The mean age was 39.9 (standard deviation, 13; range, 19–85). The university sample was skewed toward younger ages, reflecting the student population. The UPMC sample had a normal distribution curve within a typical working age range. A higher proportion of women than men responded to both surveys. Pitt statistics on enrollment of graduate and professional students by gender as of October 2008 and for faculty by gender in 2007 [50], the most recent located at the time of analysis, were compared to survey respondent data. Female students were overrepresented by 12.6% and female faculty members by 6.3%. A profile of the UPMC workforce as a whole was not readily available. Respondents' confidence in their computer skills was high (Table 1, online only).

All user groups were represented in the pool. More than half of respondents with UPMC email addresses (n=261/481, 54%) indicated that they had roles at the university as well as UPMC and were counted in both categories in Table 2 (online only). The large proportion of graduate and professional students (32%) compared to undergraduate students (7%) reflected the student mix of the health sciences schools (Table 2, online only).

4.3. Factors associated with differences in e-book use

4.3.1. Demographics and use of e-books and print books

Overall, 55.4% (n=505/911) of respondents reported using an HSLS e-book. Cross-tabulation with Pearson chi-square showed use of e-books from HSLS was not related to UPMC or university affiliation or to age of respondent, but role at their primary institution was related to use of e-books (Table 3). Over 70% of UPMC respondents in the categories of "attending physicians"; "interns, residents, or fellow"; and Pitt "postdoctoral or fellows" reported using e-books. Use of e-books by UPMC respondents in other roles ranged from 28.6% for administrators to 56.8% for researchers. At Pitt, respondents in other roles ranged from 48.9% of undergraduates to 64.7% of faculty. Gender was associated with use of e-books. Men (n=202/303, 66.7%) were more likely than women (n=279/508, 54.9%) to report using HSLS e-books ($\chi^2=10.849$, $df=1$, $P<0.001$), a statistically significant difference. In addition, there was a significant difference between the sexes in the use of e-books for in-depth study. Men (n=167/313, 53.4%) were more likely than women (n=203/551, 36.8%) to report using HSLS e-books for in-depth reading ($\chi^2=22.284$, $df=2$, $P=0.000$).

Table 3. Reported e-book use by role at University of Pittsburgh Medical Center (UPMC) or University of Pittsburgh

Affiliation and role*	Reported use of e-books	
	n	(%)
UPMC (n=435)†		
Intern, resident, or fellow (n=91)	73	(80.2%)
Attending physician (n=71)	52	(73.2%)
Researcher (n=74)	42	(56.8%)
Other (n=25)	14	(56.0%)
Other patient care (n=35)	19	(54.3%)
Support staff (n=43)	18	(41.9%)
Nurse (n=68)	28	(41.2%)
Administrator (n=28)	8	(28.6%)
University of Pittsburgh (n=648)‡		
Postdoctoral or fellow (n=74)	54	(73.0%)
Faculty or teaching role (n=215)	139	(64.7%)
Graduate or medical student (n=205)	127	(62.0%)
Staff (n=84)	43	(51.2%)
Undergraduate (n=45)	22	(48.9%)
Other (n=25)	12	(48.0%)

* Respondents can appear in more than one category. Respondents with UPMC email addresses indicating roles at the university are included in both categories.
† $\chi^2=48.051$, $df=7$, $P=0.000$.
‡ $\chi^2=13.705$, $df=5$, $P=0.018$

4.4. Barriers and facilitators to e-book use

4.4.1. Respondent awareness and use of the e-book collection

Most respondents (n=599/914, 65.5%) recalled seeing information about e-books on the HSLS website, although slightly fewer l(n=505/911, 55.4%) reported using an HSLS e-book. Use of e-books to look up brief factual information was reported by 56.6% (n=516/911), while use for in-depth study was reported by 41.9% (n=383/913).

4.4.2. Use and rating of e-book search tools

The utility of the 5 HSLS e-book search tools, Google Books, and the Amazon Search Within the Book feature was rated by 863 respondents, as summarized in Figure 1. The federated full-text search tool was used by the largest percent of respondents (n=580/863, 67.2%) and was rated moderately to extremely useful by 74.3% (n=431/580) who used it. Google Books was also rated as moderately to extremely useful by 74.3% (n=373/502) who used it. They gave the lowest ratings to the library catalog (PITTCat), with 61.2% (n=306/500) rating it moderately to extremely useful.

4.4.3. Respondent use of physical and virtual libraries

Respondents reported using the HSLS website more than they used physical libraries to answer health sciences–related questions, but 66.9% (n=617/922) indicated they used both. The HSLS website was used by 95.4% (n=883/926) in the previous month, while walk-in use of the physical library in the past month was reported by 63.8% (n=406/636). A library was the primary work or study reading place for 5.4% (n=50/925) of respondents, while 45.1% (n=417/925) read at work and 45.6% (n=422/925) read at home.

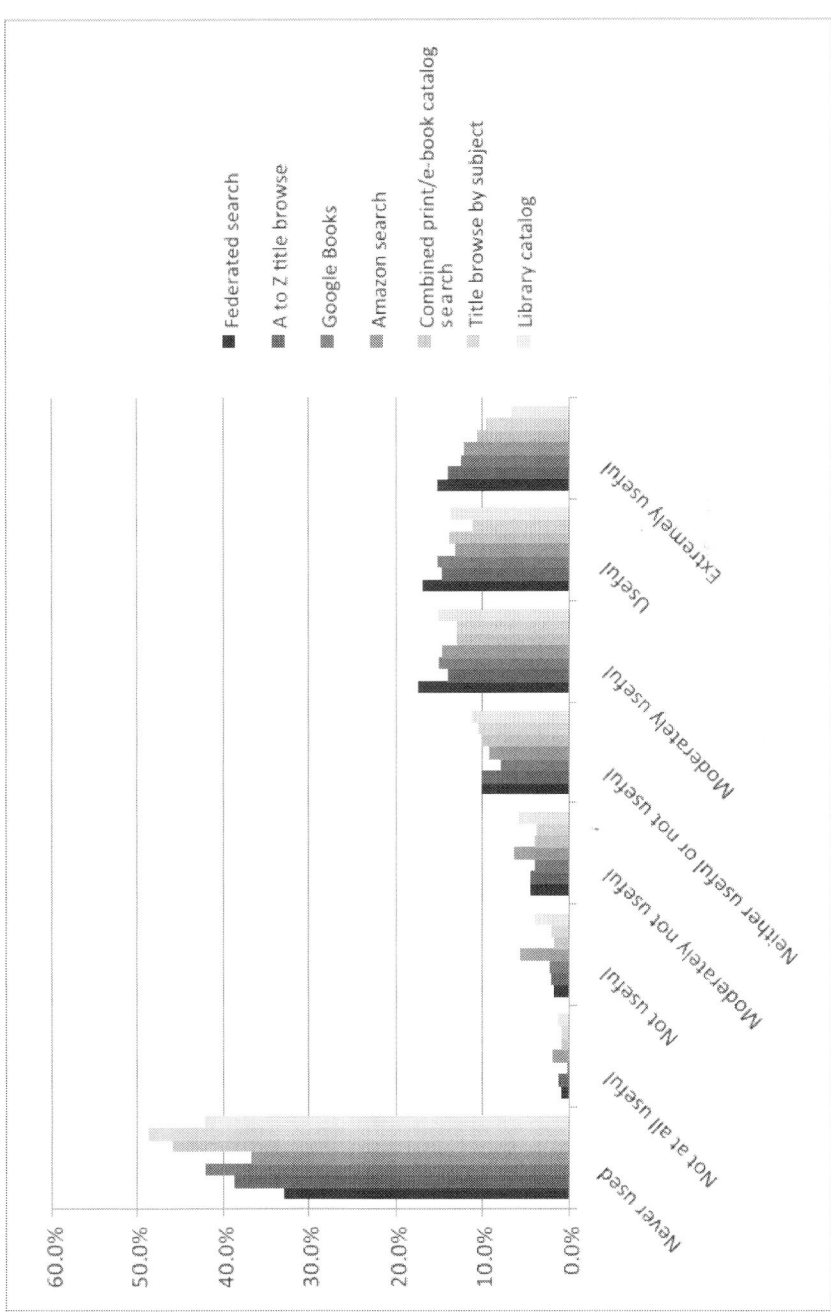

Figure 1. Utility of e-book search tools (n=863)

Of those who used a library, 67.2% (n=432/643) borrowed or used an HSLS print book in the past year. Use of print and e-books was positively related ($\chi^2=19.365$, $df=1$, $P=0.000$). The proportion using both was 44.7% (n=262/586), while the proportion using neither was 17.1% (n=100/586). Print-only use was reported by 23.4% (n=137/586), and 14.8% (n=87/586) used only e-books. Respondents reported more confidence in their ability to find the print books in their library collection than the e-books. For e-books, 46.4% (n=417/899) agreed or completely agreed that they could locate them, while 66.7% (n=610/914) expressed the same degree of confidence for locating print books. When asked if e-books were accessible where they needed to use them, 45.3% (n=406/897) agreed, and about the same proportion agreed that the print collection contained books they needed (n=412/913, 45.1%). However, only 27.9% (n=255/914) agreed or completely agreed that they had time to get a print book when they needed it.

4.4.4. Distance, time, and use of e-books and print books

Correlations were examined between distance from the library, perceived available time to get books, and use of print and electronic books (Figure 2). For those who used a physical library, the closer they worked to the library, the more likely they were to have entered the library in the past month, but distance had no significant effect on use of print library books. In all distance categories, from "in same building" (n=110/154, 71.4%) to "farther than 5 blocks away" (n=78/128, 60.9%), the majority reported using an HSLS print book in the past year ($\chi^2=6.555$, $df=3$, $P=0.088$).

4. RESULTS

* Includes only respondents who answered all questions

Figure 2. Percent using e-books and print within distance, time categories (n=558*)

The perception of lack of time to go to the library to get a book was more influential than distance on print book use. Of the respondents who agreed or completely agreed that they had time to go to the library to get a book, 84.3% (n=172/204) had used an HSLS print book in the past year, while 55.3% (n=126/228) who disagreed or completely disagreed with the statement had used an HSLS print book ($\chi^2=49.668$, $df=4$, $P=0.000$).

Available time had less effect on e-book than print book use. Of those who agreed or completely agreed that they had time to go to the library, 64.7% (n=132/204) had used an HSLS e-book in the past year, while 55.3% (n=126/228) who disagreed or completely disagreed had used an e-book ($\chi^2=5.750$, $df=4$, $P=0.219$). Distance to the library and e-book use were inversely proportional. Sixty-seven percent (n=198/296) who were located within 1 block of the library had used an HSLS e-book, while 52.3% (n=137/262) who were 2 or more blocks away had used 1 ($\chi^2=12.478$, $df=3$, $P=0.005$).

4.4.5. Importance of e-book features to users

Respondents valued printing, saving, and searching e-books more than bookmarking, highlighting, and annotating content (Figure 3). Printing was rated moderately to extremely important by 76.6% (n=661/863), while saving to a computer was given the same rating by 72.0% (n=621/863). Full-text searching was moderately to extremely important to 73.9% (n=638/863). Respondents were less interested in bookmarking, highlighting, and annotating text.

4. RESULTS

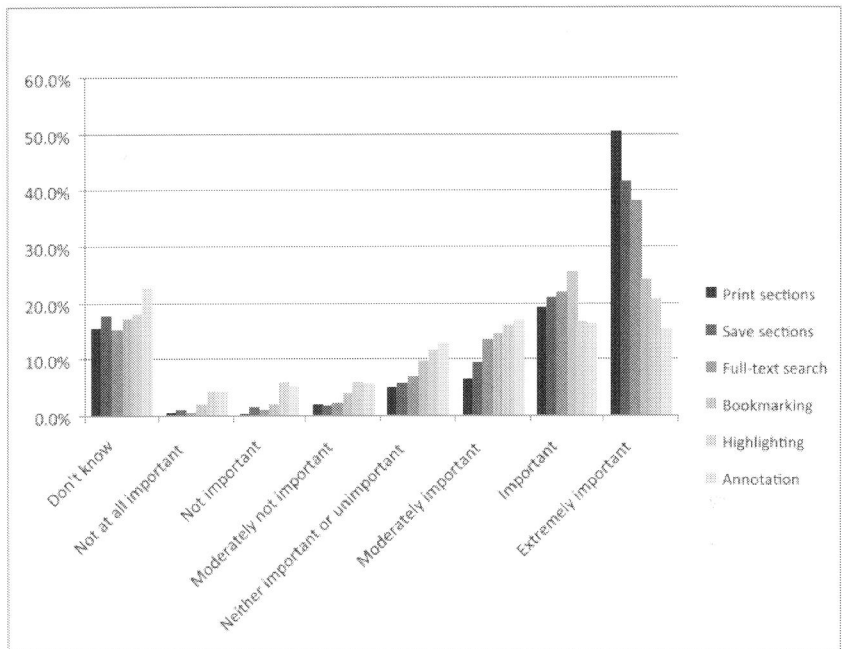

* Includes respondents who never used an Health Sciences Library System (HSLS) e-book. Results did not change when restricted to only e-book.

Figure 3. Importance of e-book features (n=863)*

4.4.6. Type of book and format preference

There were differences in format preference by type of book, as summarized in Figure 4. E-books were preferred most often for general reference and pharmaceutical reference, while print books were preferred most often for textbooks and handbooks. For all book types, some respondents were inflexible in their preference, but for each type, a large proportion (62.4%–78.7%) said they would use the format that was most convenient at time of use. Those preferring print were more flexible about using e-books than those preferring e-books were about using print.

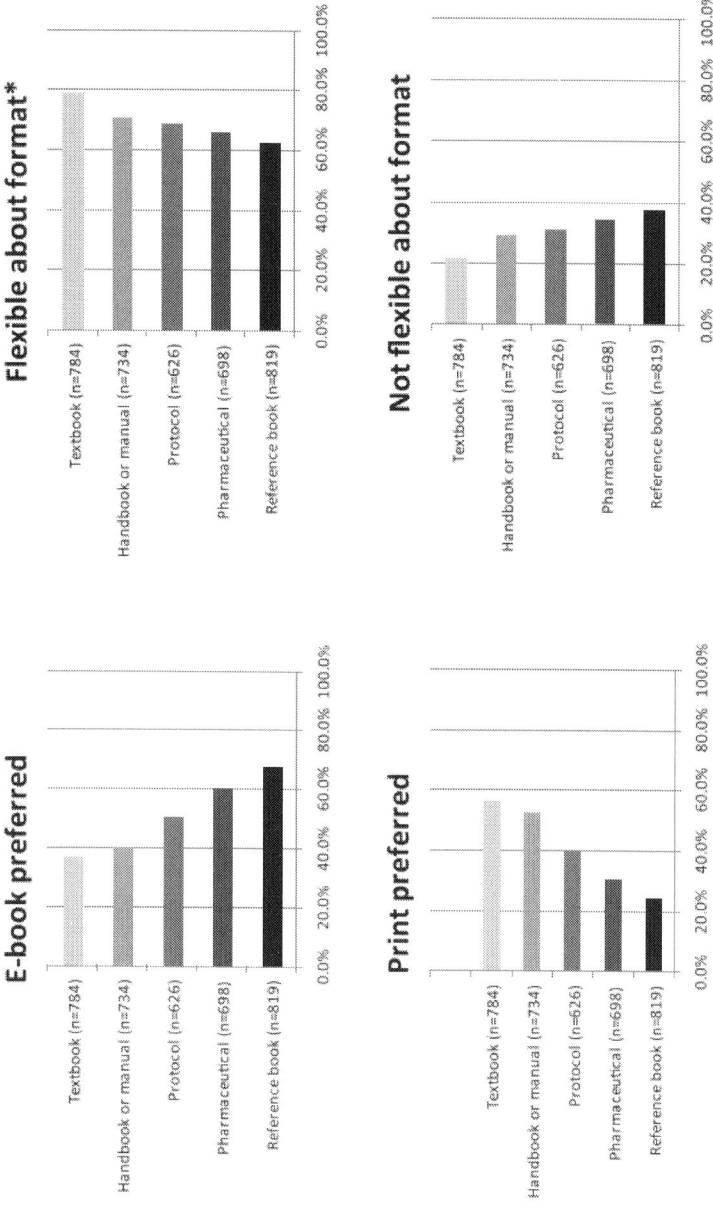

* Respondents that indicated they would use their least preferred format if it was more convenient at the time of use or indicated no preference are coded as flexible.

Figure 4. Book format preferences and flexibility*

4.5. Intended use and e-book or print preference

4.5.1. E-book use for clinical care, teaching, learning, and research

The survey indicates that UPMC respondents are using e-books. Their intended use varied by job category. E-books were used for clinical care by 75.3% (n=55/73) of attending physicians; 86.0% (n=86/100) of interns, residents, and fellows; and 38.9% (n=28/72) of nurses. They were used by 61.8% (n=21/34) of other clinical care specialists, such as respiratory care and physical therapists. Almost half of UPMC administrators (n=14/30, 46.7%) reported using e-books to support administrative tasks.

At the university, 76.5% (n=62/81) of postdoctoral students and fellows, and 54.1% (n=124/229) of faculty reported using e-books to support research. E-books were less frequently assigned for class readings. Only 21.3% (n=37/174) of people with teaching responsibilities reported assigning a class reading from an e-book. Fewer undergraduate students (n=7/50, 14.0%) than graduate and medical students (n=77/230, 33.5%) reported being assigned a class reading from an e-book, while 51.0% (n=25/49) of undergraduates and 62.1% (n=139/224) of graduate and medical students used an e-book to complete an assignment.

4.6. Limitations of the study

The study had a large enough sample to detect differences in e-book use by various academic health sciences library user groups, and the mix of survey respondents reflected the mix of roles and institutional affiliations at the university and hospitals. However, the response rate to the survey was lower than expected, despite the use of several methods to increase response rates, including a cover email from the library director endorsing the survey and stating its importance to library users and several follow-up reminder emails to nonresponders. This suggests that nonresponse bias could apply. The sampling frame, library users with remote access passwords, might also have biased the results, possibly overestimating the proportion of all library patrons who used e-books. Other possible limitations to the study included the absence of complete data from participants who did not finish the survey (n=108/979, 11.0%) and potential confusion over the definition of e-books. While a working definition was given at the beginning of the survey, information in the comments to the survey indicated that several participants were unsure of what an e-book was. Finally, the results might not be generalizable to other libraries with different user populations and collections.

5. DISCUSSION

Information need drives both e-book and print use, with contextual factors such as distance from the library less important determinants of use. The volume of e-book use by library patrons varied according to their different roles, reflecting the information-intensive qualities of those roles, with students, postdoctoral fellows, researchers, and clinical physicians among the heaviest e-book users. A high volume of e-book use was also associated with a high volume of print book use. Some of the heaviest users of the e-book collection were within one block of a library, disproving the intuitive idea that e-book use would increase with distance from the library.

Preference for e-books or print varied with the type of book. Study respondents' preferences indicated that reference books or pharmaceutical references were the best candidates for e-books. Most surprisingly, a large percentage of users (62.4%–78.7%) claimed they were flexible with respect to print or electronic format, stating that they would use their least preferred format if it were the most convenient to access at the time of need. This should give collection development librarians more confidence that purchasing reference books and other

5. DISCUSSION

essential medical books that are likely to be used like reference books [32, 33, 39] as e-books will satisfy the majority of users' information needs, especially if more e-book vendors offer printing of whole chapters for offline in-depth study. If economics dictate that duplication of books in both formats must be reduced, a combination of promoting the e-book collection to increase awareness and educating users to increase user e-book skills may increase the adoption of e-books by those who prefer print.

Awareness of HSLS e-books (n=599/914, 65.5%) was comparable to that in other studies [30, 33, 35, 38]. The survey itself promoted e-book awareness: 17 respondents said in free-text comments that they were unaware of HSLS e-books before taking the survey. Others called for increased promotion and educational efforts, such as the respondent who said, "I wish they were better 'advertised' as available resources. I kind of happily stumbled upon them. It would be great if it were more widely known." This is good news, as HSLS designed the website so that users could stumble on resources without librarian intervention, but supplementing good web design with active promotion—such as inclusion in library orientations, newsletter articles, and in-person or online training sessions, as recommended by Dinkelman and Stacy-Bates—could increase use [19]. User education may be key to expanding use of the e-book collection. As one respondent said:

The librarians at [UPMC] Shadyside have helped me learn how to easily access the HSLS in the past two years. I was surprised that after doing this survey my preference has now changed to wanting to use an electronic source. It is all because I now know how to easily find what I want in the HSLS.

Study findings support claims that users prefer web access to e-books over library catalog access [5, 27, 32,33, 35, 42]. Prominence on the HSLS home page and enhanced access to e-book content may be responsible for the high approval rating for the HSLS federated e-book search. It was preferred over the library catalog by survey respondents. While some commented that the federated e-book search was slow, they appreciated the enhanced access to the content of the e-book collection that it provides. However, users might not know the federated search did not include all e-books available at the university. The federated search tool represents a step in the right direction, but more inclusive full-text search options across all Pitt e-book collections would increase access to information. Since the survey, the "A to Z" e-book title list has been removed from the website. Even though survey respondents rated it highly, HSLS librarians felt that as the e-book collection expanded, this list became too long for useful browsing. No complaints were received following the "A to Z" list removal, perhaps because the four remaining e-book search tools are sufficient.

Users indicated that they would be willing to use a less preferred format, if it were more convenient at the time of need. The development of a more sophisticated and prominently placed combined e-book and print search tool should allow users to more easily discover all the available format options. They rated the library catalog lower than other search tools, yet only by searching the catalog can they locate the full range of e-book titles available to them at Pitt. Users rated Google Books, which provides a combination of full-text searching with catalog access for print book location, very highly. If the Google Books interface could be extended to reliably link library patrons to e-books in their library, perhaps this would be the most useful e-book discovery tool.

The HSLS e-book collection is heavily used for clinical, research, and individual study purposes. This result mirrors those of other studies [27, 30]. That attending physicians, medical students, postgraduate medical trainees, and researchers used e-books most heavily was not surprising, given the information-intensive nature of those roles. One physician respondent referred to e-books as a "lifesaver in my clinical position." Some respondents commented that internal medicine and surgery subjects were well represented in the collection, but pediatrics and pathology were not, and that they wished there were more e-books in their specialties. This indicates some user groups may not be using the e-books because the collection does not include what they want. Collection analysis and focus group discussions with representatives of different user groups could help identify strengths and weaknesses in the collection by topic area and specialty.

The HSLS e-book collection was not used heavily by faculty for assigning class readings but was used frequently by students to complete course assignments. E-book chapters cannot be easily posted to course management software, and licensing of sufficient simultaneous e-book users for class access purposes can be problematic. If e-book publishers want their products to support classroom use, they should consider "journalizing" them as SpringerLink has done, allowing saving of chapters and posting to course management systems [5]. This would also increase their appeal to users who want to print or save chapters for future use.

6. CONCLUSION

Moving forward, librarians should consider several courses of action based on the study results. First, while passive promotion through cataloging and prominent placement on the library website brings e-books to the attention of many library users, more active instruction and promotion is needed to increase use of the collection by library patrons. Second, because patrons prefer Internet access to library catalog access, every effort should be made to ensure that e-book catalog records can be repurposed for web access with minimal technical effort and that, whenever possible, full-text search options are provided to enhance access to book content. Finally, user flexibility about book format indicates collection development librarians can selectively reduce duplication of titles in print and electronic forms.

This study looked at e-book use in relation to many factors—demographic, affiliation, reason for use, and type of book—and found that health sciences users are flexible about what they use and will get the information they need, however they need to get it. Perhaps librarians are spending too much time thinking about information containers (print versus electronic), a library-centric way of thinking, and not about the content. Study respondents were frequent users of information, using it in whatever container it comes in and locating it with a variety of access methods.

ACKNOWLEDGMENTS

The authors thank Dr. Don Musa, University of Pittsburgh, University Center for Social and Urban Research, for his assistance with developing and testing the survey instrument.

REFERENCES

1. Tannery N.H, Foust J.E, Gregg A.L, Hartman L.M, Kuller A.B, Worona P, Tulsky A.A. Use of web-based library resources by medical students in community and ambulatory settings. J Med Libr Assoc. 2002 Jul;90(3):305–9.

2. D'Alessandro M.P, D'Alessandro D.M, Galvin J.R, Erkonen W.E. Evaluating overall usage of a digital health sciences library. Bull Med Libr Assoc. 1998 Oct;86(4):602–9.

3. De Groote S.L, Dorsch J.L. Measuring use patterns of online journals and databases. J Med Libr Assoc.2003 Apr;91(2):231–41.

REFERENCES

4. De Groote S.L, Shultz M, Doranski M. Online journals' impact on the citation patterns of medical faculty. J Med Libr Assoc. 2005 Apr;93(2):223–8.

5. van der Velde W, Ernst O. The future of ebooks? will print disappear? an end-user perspective. Libr Hi Tech. 2009;27(4):570–83.

6. Foust J.E, Bergen P, Maxeiner G.L, Pawlowski P.N. Improving e-book access via a library-developed full-text search tool. J Med Libr Assoc. 2007 Jan;95(1):40–5.

7. MacCall S.L. Online medical books: their availability and an assessment of how health sciences libraries provide access on their public websites. J Med Libr Assoc. 2006 Jan;94(1):75–80.

8. Association of Academic Health Sciences Libraries. Annual statistics of medical school libraries in the United States and Canada, 2003–2004, 2006–2009 [database]. 27th, 30th–32nd ed. Seattle, WA: The Association;

9. Ugaz A.G, Resnick T. Assessing print and electronic use of reference/core medical textbooks. J Med Libr Assoc. 2008 Apr;96(2):145–7.

10. Koestner B.A. Evaluating a medical library's print and electronic book collection: the balanced scorecard approach [master of science in library science ed.] [Internet] Chapel Hill, NC: University of North Carolina at Chapel Hill; 2009 [cited 9 Jul 2010]. < http://ils.unc.edu/MSpapers/3500.pdf>.

11. Shedlock J. The future of books at the Galter Health Sciences Library, Feinberg School of Medicine, Northwestern University [Internet] Doody's; 2010 [cited 7 Jul 2010]. <http://www.doody.com/dct/PublicFeaturedArticle.asp?SiteContentID=23>.

12. Newman M. 2009 librarian ebook survey [Internet] Palo Alto, CA: HighWire Press, Stanford University; 2009 [cited 9 Jul 2010]. < http://highwire.stanford.edu/PR/HighWireEBookSurvey2010.pdf>.

13. Fowler G.A, Worthen B. Amazon to launch Kindle for textbooks. Wall Street J [Internet] 2009 May 5 [cited 29 Nov 2010]. < http://online.wsj.com/article/SB124146996831184563.html>.

14. Behler A. E-readers in action: an academic library teams with Sony to assess the technology. Am Libr.2009 Oct;40(10):56–9.

15. McClure M. Turning a new page in ebooks. Inf Today. 2009 Apr;26(4):1–18.

16. Kiernan V. An ambitious plan to sell electronic books. Chron High Educ. 1999 Apr;45(32):A29.

17. Dillon D. E-books: the University of Texas experience, part 2. Libr Hi Tech. 2001;19(4):350–62.

18. Rowlands I, Nicholas D. Understanding information behaviour: how do students and faculty find books.J Acad Libr. 2008 Jan;34(1):3–15.

19. Dinkelman A, Stacy-Bates K. Accessing e-books through academic library web sites. Coll Res Libr.2007 Jan;68(1):45–58.

20. Belanger J. Cataloguing e-books in UK higher education libraries: report of a survey. Program: Electron Libr Inform Syst. 2007;41(3):203–16.

21. Armstrong C, Lonsdale R. Challenges in managing e-books collections in UK academic libraries. Libr Collect Acquis Tech Serv. 2005 Spring;29(1):33–50.

22. Dillon D. E-books: the University of Texas experience, part 1. Libr Hi Tech. 2001;19(2):113–24.

REFERENCES

23. Littman J, Connaway L.S. A circulation analysis of print books and e-books in an academic research library. Libr Res Tech Serv. 2004;48(4):256–62.

24. Grigg K.S, Koestner B.A, Peterson R.A, Thibodeau P.L. Data-driven collection management: through crisis emerge opportunities. J Electron Resour Med Libr. 2010 Mar;7(1):1–12.

25. Sprague N, Hunter B. Assessing e-books: taking a closer look at e-book statistics. Libr Collect Acquis Tech Serv. 2008 Sep;32(3–4):150–7.

26. Peterson M.W, Rowat J, Kreiter C, Mandel J. Medical students' use of information resources: is the digital age dawning. Acad Med. 2004 Jan;79(1):89–95.

27. ebrary 2007 global faculty e-book survey [Internet] 2007 [cited 10 Jun 2009]. <http://www.ebrary.com/corp/ collateral/ en/ Survey/ ebrary_faculty_survey_2007.pdf>.

28. Grudzien P, Casey A.M. Do off-campus students use e-books. J Libr Admin. 2008 10;48(3):455–66.

29. Raynor M, Iggulden H. Online anatomy and physiology: piloting the use of an anatomy and physiology e-book-VLE hybrid in pre-registration and post-qualifying nursing programmes at the University of Salford.Health Inf Libr J. 2008 Jun;25(2):98–105.

30. Shelburne W.A. E-book usage in an academic library: user attitudes and behaviors. Libr Collect Acquis Tech Serv. 2009;33(2–3):59–72.

31. Rowlands I. Superbook: planning for the ebook revolution [Internet] 2007 [cited 2 Feb 2009]. <http://www.ucl.ac.uk/ infostudies/research/ciber/superbook/>.

32. Nicholas D, Rowlands I, Clark D, Huntington P, Jamali H.R, Olle C. UK scholarly e-book usage: a landmark survey. ASLIB Proc. 2008;60(4):311–34.

33. Levine-Clark M. Electronic book usage: a survey at the University of Denver. Portal. 2006 Jul;6(3):285–99.

34. Franklin B, Plum T. Library usage patterns in the electronic information environment. Inf Res [Internet]2004;9(4):paper 187 [cited 4 Nov 2009]. < http://informationr.net/ir/9-4/paper187.html>.

35. ebrary 2008 global student e-book survey [Internet] 2008 [cited 22 Feb 2009]. <http://www.ebrary.com/corp/ collateral/en/ Survey/ebrary_student_survey_2008.pdf>.

36. Chong P.F, Lim Y.P, Ling S.W. On the design preferences for ebooks. IETE Tech Rev. 2009 May–Jun;26(3):213–22.

37. Levine-Clark M. Electronic books and the humanities: a survey at the University of Denver. Collection Building. 2007;26(1):7–14.

38. Abdullah N, Gibb F. Students' attitudes towards e-books in a Scottish higher education institute: part 1.Libr Rev. 2008;57(8):593, 605.

39. Strother E.A, Brunet D.P, Bates M.L, Gallo J.R., 3rd Dental students' attitudes towards digital textbooks.J Dent Educ. 2009 Dec;73(12):1361–5.

40. Appleton L. The use of electronic books in midwifery education: the student perspective. Health Info Libr J. 2004 Dec;21(4):245–52.

41. Hernon P, Hopper R, Leach M.R, Saunders L.L, Zhang J. E-book use by students: undergraduates in economics, literature, and nursing. J Acad Libr. 2007 Jan;33(1):3–13.

42. Rowlands I, Nicholas D, Jamali H.R, Huntington P. What do faculty and students really think about e-books. ASLIB Proc. 2007; 59(6):489–511.

REFERENCES

43. Liu Z, Yang Z.Y. Factors influencing distance-education graduate students' use of information sources: a user study. J Acad Libr. 2004 Jan;30(1):24–35.

44. Jamali H.R, Nicholas D, Rowlands I. Scholarly e-books: the views of 16,000 academics results from the JISC national e-book observatory. ASLIB Proc. 2009;61(1):33–47.

45. Appleton L. Using electronic textbooks: promoting, placing and embedding. Electronic Libr.2005;23(1):54–63.

46. Aday L.A, Cornelius L.J. Deciding how many will be in the sample. In: Designing and conducting health surveys: a comprehensive guide. 3rd ed. San Francisco, CA: Jossey-Bass; 2006. pp. 154–93. p.

47. American Association for Public Opinion Research. Standard definitions: final dispositions of case codes and outcome rates for surveys [Internet] Lenexa, KS: The Association; 2009 [cited 20 Apr 2009]. <http://www.aapor.org/Content/ NavigationMenu/ Resources forResearchers/StandardDefinitions/StandardDefinitions2009new.pdf>.

48. Aday L.A, Cornelius L.J. Choosing methods of data collection. In: Designing and conducting health surveys: a comprehensive guide. 3rd ed. San Francisco, CA: Jossey-Bass; 2006. pp. 100–23. p.

49. Shih T, Fan X. Comparing response rates from web and mail surveys: a meta-analysis. Field Methods.2008 Aug;20(3):249–71.

50. University of Pittsburgh, Office of Institutional Research. Common data set 2007–2008 [Internet]Pittsburgh, PA: The University [cited 30 Mar 2009]; <http://www.ir.pitt.edu/cds/ documents/ Pittsburgh CDS_024.pdf>.

CHAPTER 7

Knowledge and skills for the digital era academic library

J. Raju

Library and Information Studies Centre, University of Cape Town, Private Bag X3, Rondebosch, 7701 South Africa

ABSTRACT

Technology has altered the traditional academic library beyond recognition. These dramatic changes have impacted significantly on the knowledge and skills requirements for LIS professionals practising in this environment. While there have been studies in other parts of the world which have investigated the knowledge and skills requirements for the digital era academic library environment, to date no comprehensive study has 'drilled' down into this area in the South African context. This paper reports on a preliminary study which is part of a wider study aimed at developing a comprehensive skills statement which would provide an objective framework against which professional LIS practitioners in the modern academic library environment in South Africa may both measure their existing competencies and also identify the need for further skills acquisition. The research question guiding this preliminary investigation was: What key knowledge and skills are required for LIS professionals to effectively and efficiently practise in a digital era academic library in South Africa? The triangulated findings (using content analysis of job advertisements and semi-structured interviews) from this preliminary investigation are used to ascertain an initial picture of key knowledge and skills sets required for LIS professionals in this environment. These preliminary findings also proved useful in teasing out some of the parameters for the wider study targeting the development of a comprehensive skills statement for higher education libraries in South Africa. The study reported here has relevance for the academic library context in other parts of the world as well.

KEYWORDS

Academic libraries; Digital era; Disciplinary knowledge; Generic skills; Job advertisements; Personal competencies

1. INTRODUCTION

New methods of scholarly communication, expansion of the library's virtual space via knowledge or research commons, the proliferation of social media, and the explosive growth of mobile devices, tablets and related applications, have collectively altered the traditional academic library beyond recognition. These dramatic changes, largely the result of rapidly evolving information and communication technologies (ICTs), have impacted significantly on the knowledge and skills requirements for library and information science (LIS) professionals practising in this environment. The transformed landscape requires a new generation of LIS professionals to effectively and efficiently mediate it. Orme (2008, p. 627–628) categorised knowledge and skills required for this transformed environment into: discipline-specific knowledge (that is, knowledge that

1. Introduction

relates specifically to the LIS profession), generic skills (general skills which apply to all disciplines) and personal competencies (attitudes, values and personal traits). Choi and Rasmussen (2009, p. 465), through content analysis of job advertisements in the United States of America (USA), found that key disciplinary knowledge required for this digitally oriented environment included understanding metadata, and knowledge and experience in digital content creation and management. Generic skills such as effective communication and interpersonal skills, critical thinking, problem solving and teamwork were found by Nonthacumjane (2011, p. 283) to be required by information professionals in a digital library environment in both Norway and Thailand. Howard's (2009) Australian master's study (cited in Nonthacumjane, 2011, p. 283) highlighted personal competencies such as flexibility, adaptability and reflective thinking as being required for working in a digital library environment. What about knowledge and skills requirements for the digital era South African higher education academic library environment? To date, no comprehensive study has 'drilled' down into this area in the South African context. South Africa is richly endowed with academic libraries. These are located in the country's 23 universities. Many of these higher education libraries offer state-of-the art LIS services to academic and research communities in universities in the country which are leading institutions on the African continent. Hence an analysis of knowledge and skills requirements for the digital era academic library environment in South Africa is important.

This paper reports on a preliminary study which is part of a wider study aimed at developing a comprehensive skills statement which would provide an objective framework against which professional LIS practitioners in the modern academic library environment in South Africa may both measure their existing competencies and also identify the need for further skills acquisition. Such a skills statement would also be useful in informing curriculum review and revision in LIS education and training, as academic libraries in South Africa are a major employer of LIS graduates (Ocholla & Shongwe, 2013, p. 38). There are ten LIS schools in South Africa, based in universities, which offer education and training in response to the knowledge and skills requirements of the LIS work environment. The research question guiding this preliminary investigation was: What key knowledge and skills are required for LIS professionals to effectively and efficiently practise in a digital era academic library in South Africa? The triangulated findings (using content analysis of job advertisements and semi-structured interviews) from this preliminary investigation are used to ascertain an initial picture of key knowledge and skills sets required for LIS professionals in this environment. These preliminary findings also proved useful in teasing out some of the parameters for the wider study targeting the development of a comprehensive skills statement for higher education libraries in South Africa. While these early findings draw from South African higher education libraries, the study is contextualized within global trends affecting higher education libraries and hence has relevance for other parts of the world as well.

2. LITERATURE REVIEW

A review of pertinent literature locates this study in the global context of the changing academic library and its impact on knowledge and skills requirements for academic libraries in the digital era.

2.1. The academic library in the digital era

As technology continues to impact on the delivery of information services, traditional academic library systems, point out Choi and Rasmussen, 2006 and Choi and Rasmussen, 2009, have come to embrace the digital library model. Academic libraries have evolved from focusing on the management of physical resources and related services to "transforming resources and services into digital formats to support teaching, learning and research" (Choi & Rasmussen, 2009, p. 457). O'Connor and Au (2008, p. 57) too state that the academic library has changed dramatically and with this change the "conception of a digital library has become a reality".

Changes in areas of teaching and learning, influenced and enabled by technology, remarks McCarthy (2005), have impacted on academic libraries — for example, the creation of new knowledge products such as subject portals and subject specific websites to support teaching and learning; or the re-purposing of physical spaces and the expansion of virtual spaces to support new pedagogies and changes in the teaching and learning process.

In terms of research, Luce (2008) explains that the "convergence of exponential increases in computing, storage, online sensors, and bandwidth" has enabled scientists and researchers to collaborate in new ways thus leading to the rise of eScience and eResearch. eScience developments now characterise not only the Science, Technology, Engineering and Medicine disciplines but the Humanities and Social Sciences as well. The convergence of technologies has led to "new ways of thinking about and understanding physical, biological and social phenomena" (Luce, 2008). These revolutionary developments have demanded an equally dramatic shift in the way academic or research libraries serve the needs of scientists in the new eScience and eResearch contexts.

Luce's (2008) advice to academic libraries which are positioning themselves to support eResearch, is to be cognizant of the fact that knowledge preservation becomes one of the key roles of such a library. Luce (2008) goes on to emphasize that this knowledge preservation necessitates:

- Ensuring the quality, integrity, and curation of digital research information;
- Sustaining today's evolving digital service environments;

- Bridging and connecting different worlds, disciplines, and paradigms for knowing and understanding; and

- Archiving research data in a data world.

Hence to enable digital capture, curation, preservation and sharing of knowledge, the academic library in the digital era needs to reflect a service environment that embraces digitization, electronic publishing, Web 2.0, Web 3.0, Library 2.0, Library 3.0, social media, open access, and a host of other fast evolving ICTs. As academic libraries the world over shift into this digital era, these developments and innovations impact on the knowledge and skills profiles of LIS professionals in academic libraries (Choi and Rasmussen, 2009 and Nonthacumjane, 2011). New skills sets are required to mediate this digitally oriented academic library environment.

2.2. Knowledge and skills requirements in the digital academic library environment

A study of job advertisements by Orme (2008, p. 630) found that "a mixture" of discipline-specific knowledge (also referred to as professional knowledge), generic skills and personal competencies is required of LIS professionals in a digitally oriented LIS environment. Nonthacumjane too (2011, p. 280, p. 286), in a study that probed the "skills and competencies required for LIS professionals to be effective and efficient working in the digital era", concludes that they should be able to dynamically exercise "personal, generic and discipline-specific skills".

2.3. Disciplinary knowledge

Partridge and Hallam (2004) use the "double helix image of human DNA" to argue that both disciplinary knowledge and generic capabilities

2. Literature review

"make up the genome of the successful information professional in the information age". Many have argued that the new digitally oriented academic library is a renewed conceptualisation of traditional LIS resources and services, now undergirded and driven by new technologies (Choi and Rasmussen, 2009, Gerolimos and Konsta, 2008, McCarthy, 2005, Middleton, 2003 and Missingham, 2006). Hence in a digital academic library environment core knowledge and skills of traditional librarianship are important but need to be augmented by new technological knowledge (Choi & Rasmussen, 2009, p. 465).

Tammaro (2007, p. 237) cogently points out that

> *cataloguing and classification skills have much relevance to the Web…a more thorough knowledge of the major schemes and their working principles is required to allow a person to adapt and accommodate existing metadata schemes to use, and to possess the basic expertise to construct new schemes.*

Digital library applications are closely linked to Web technology (Choi & Rasmussen, 2009, p. 463). Consequently, as modern academic libraries move into the creation of digital content and its organisation and preservation through metadata creation and management to make their special collections more accessible via the Web, the need for knowledge of the following technologies becomes critical: digital library architecture and software, technical and quality standards, HTML coding, general computer skills and computer literacy, database development and management, Web mark-up languages such as SGML and XML, and Web development and design (Choi and Rasmussen, 2006 and Choi and Rasmussen, 2009). Ocholla and Shongwe (2013, p. 39, 42) in their content analysis of job advertisements in South Africa over a four-year period (2009–2012) found IT skills to be very sought after in libraries as "more information services", particularly in academic and research environments, become "IT or e-access and e-service dependent".

Parallel to the development of digitization and curation of unique collections, the academic library in the digital era is also being challenged by an emerging trend of research data management and curation (Wise, Henninger, & Kennan, 2011, p. 279). The emergence of eScience and eResearch is accompanied by the generation of vast amounts of research data in need of collection, preservation, management and provision for future access to enable re-using, re-purposing, re-combining, etc. According to Luce (2008) metadata is an essential component of research data and research or academic libraries, because of their traditional knowledge base, are well positioned to "lead the development of standardized, ontologically rich automated metadata" for research data sets. Here again, disciplinary knowledge, that is, the creation and management of metadata which are established tasks in the LIS profession, allows academic libraries to take responsibility for the curation and preservation of data for its re-use when needed (Luce, 2008). The Ocholla and Shongwe (2013, p. 41) study makes reference to "new job titles" emerging in the LIS job market in South Africa and puts forward that "these new titles represent strong ICT elements" and shows the influence of ICTs on emerging knowledge and skills requirements in the LIS sector. In the USA, Riley-Huff and Rholes (2011, p. 129–130, p. 135) make a similar observation in their survey of library administrators and librarians from academic libraries: "new job categories are being defined in LIS" which "while centering on core librarianship principles [knowledge organisation; knowledge dissemination; etc.]", or what this paper has been referring to as disciplinary knowledge, call for a "significant technology skills set" (a point alluded to earlier).

2.4. Generic skills

Also referred to as 'transferable skills' or 'graduate attributes' (Partridge & Hallam, 2004), generic skills refer to life skills such as communication and interpersonal skills, critical thinking, problem solving and teamwork which

2. Literature review

allow individuals to function not only in disciplinary or subject domains but also in employment and social situations. Orme's (2008, p. 626) content analysis of job advertisements in the USA revealed that from among disciplinary, generic and personal skills, "generic skills were most frequently sought", followed by disciplinary and then personal skills. Orme (2008, p. 626), however, emphasizes that while this supports the 'move to the generic' argument in the literature (Kennan et al., 2006 and Wise et al., 2011) (that is, it has become important to appoint individuals with, for example, a capacity for continuous learning and who are adaptable in a fast changing work environment rather than those who simply possess specific disciplinary knowledge), disciplinary knowledge remains important to LIS employers. This, according to Orme (2008, p. 626), is evidenced by the marked presence of disciplinary or professional skills in the top twenty most frequently sought requirements examined.

The literature reveals communication skills to be among the most highly ranked generic skills (Gerolimos and Konsta, 2008, Middleton, 2003, Orme, 2008 and Reeves and Hahn, 2010). Of particular relevance to the modern academic library is the observation that the demand for interpersonal skills is a reflection of the "reality that team-based approaches are a common practice in digital projects and interpersonal skills are a key to success in team efforts" (Choi & Rasmussen, 2009, p. 464). While general computing or computer literacy, like information literacy, would categorise as a generic skill, technology skills, for the purpose of this paper, have been discussed above under disciplinary skills because of the close alignment between digital library applications and ICTs (Riley-Huff & Rholes, 2011, p. 130). A similar principle is applicable to managerial skills, particularly in the context of the management of digitization, curation and data management projects.

2.5. Personal competencies

Choi and Rasmussen (2009, p. 457) identify from the literature the following personal attributes as being important in the LIS work environment: capacity for continuous learning, flexibility, fostering change and the capacity to work independently. To this list may be added enthusiasm and self-motivation (Orme, 2008, p. 621–622), reflective thinking, and the ability to respond to others' needs (Howard (2009) cited inNonthacumjane, 2011, p. 283). Choi and Rasmussen (2009, p. 464) highlight "adaptive skills to keep up with changes and challenges within library and information environments".

While some studies, for example Partridge and Hallam (2004), have conflated generic skills and personal attributes into a single category of generic capabilities, this paper has kept them separate for ease of analysis of data collected. It is evident from the review of literature that findings from studies of job requirements in the LIS professional sector, including the academic library sector, demonstrate the need for new generation LIS professionals to be "multi-skilled" (Orme, 2008, Reeves and Hahn, 2010 and Wise et al., 2011). In order for LIS professionals to effectively mediate the digital academic library environment, they would need to embrace a blend of discipline-specific knowledge, generic skills and personal competencies. Partridge and Hallam (2004) reiterate this: "…generic capabilities [which may include personal competencies] and discipline knowledge are quite significantly intertwined and interrelated and vital for success as a library and information professional in the twenty first century". Orme's (2008, p. 629–630) study concludes that to this "mixture" one needs to add experience as a significant number of job advertisements require experience in disciplinary and/or generic skills. Reeves and Hahn (2010, p. 115) too, in their survey of job advertisements, found that "a great many job ads requested work experience in specific functional areas of libraries…", particularly in technical service functional areas such as that requiring "cataloguing and metadata experience".

Table 1 summarises the preponderance in the literature of some of the broad trends relating to knowledge and skills requirements in the digital era academic library environment. In attempting such a table of findings from the literature, the researcher selected for inclusion in the table, only literature that:

- reported more recent studies (roughly the last 10 years, 2004–2013, but favouring more recent years);

- reported on key survey studies involving content analysis of mainly job advertisements (as the present study does) as well as literature and practitioner/employer surveys on LIS knowledge and skills requirements; and,

- included at least one recent significant South African study in the area of LIS workplace knowledge and skills requirements.

3. METHODOLOGY

The preliminary study reported here adopts a qualitative approach by using content analysis of job advertisements and semi-structured interviews with purposively selected academic library professional LIS personnel to preliminarily ascertain the knowledge and skills requirements for a modern academic library in South Africa. As mentioned at the outset, the research question guiding this preliminary investigation was: What key knowledge and skills are required for LIS professionals to effectively and efficiently practise in a digital era academic library in South Africa? Content analysis is a commonly used descriptive technique for analysing the content of a document or other communication to discover features and patterns (Neuman, 2006, p. 44). Content analysis of job advertisements is a well-established method of researching requirements of the employment market in a particular sector (Orme, 2008, p. 620, p. 623) and is very useful in reflecting the demands of employers, employment opportunities and emerging trends in the employment market.

Table 1. Knowledge and skills requirements for the digital era academic library.

3. Methodology

While '…technology associated with LIS application…' is treated as a separate trend in Table 1 (for purposes of honing in on this significant trend), in other parts of this paper, as explained earlier, it is discussed under disciplinary skills due to the close association between digital library applications and ICTs. Table 1 also provides percentages (calculated against the 11 core research papers selected by the researcher for inclusion in the table) reflecting relative preponderance in the literature of the identified trends. Table 1 proves useful, in the Discussion section of this paper, in effecting comparisons between the findings of the preliminary study being reported and that in the literature reviewed in this paper.

In view of the fact that this study was a precursor to an envisaged wider study, the *Mail & Guardian*, a weekly newspaper known nationally for carrying advertisements for vacancies in the higher education sector in South Africa, was initially used as a single source of advertisements for professional LIS posts (that is, posts requiring a professional LIS qualification) in academic libraries in South Africa. The researcher (towards the end of 2012) attempted to source relevant advertisements from the popular 'Jobs and Academic' section of the newspaper made available to the researcher at the newspaper's Cape Town satellite office. The researcher began searching, retrospectively, for advertisements from 2008 onwards on the assumption that this would deliver a significant number of 'recent' advertisements for LIS professional positions in academic libraries.

The *Mail & Guardian* did not have a publically available electronic means of searching its content — hence the retrospective searching had to be done manually. The researcher did not have access to issues of the newspaper that were at the binders at the time of searching. Added to this was the fact that a more recent trend is that many institutions do not place the full advertisement in the press — often an abbreviated form, largely for reasons of costs, appears with a URL to access the full advertisement which is removed from the relevant website once the closing date has passed, thus 'blocking' access to important research data. Reeves and Hahn (2010, p, 118) also allude to this difficulty: "… job ads typically appear now online, …a distinct advantage for job-seekers, but a serious problem for future studies of this type if the ads are not captured and stored before they disappear". The net result of the retrospective searching for advertisements was that only 39 were found. The researcher supplemented this by searching differently for 2013 advertisements. The *Mail & Guardian* was purchased on a weekly basis and scanned immediately for relevant advertisements, including searching websites referred to by abbreviated advertisements. Over and above this, *LiasaOnline*, which is the listserv of the Library and Information Association of South Africa (LIASA), was accessed for relevant advertisements, following the advice that "…online job ads are essential to getting a comprehensive picture of the job market (Reeves & Hahn, 2010, p. 105)". This approach for the whole of 2013 yielded a further 32 advertisements bringing the total to 71 (excluding duplicate) advertisements.

To triangulate findings from the job advertisements for this exploratory exercise, the researcher purposively selected, for conducting semi-structured interviews, three professional LIS staff members each (six interviews in total) from the academic libraries of the University of Cape Town and Stellenbosch University, two of South Africa's research-led universities with world rankings — an obvious indicator of their state-of-the art academic libraries. It was also the intention of the researcher to use this modest number of interviews to trial this particular method of data collection for this type of study.

The interviewees in the purposive sampling were selected based on the researcher's knowledge and research experience (Welman, Kruger, & Mitchell, 2005, p. 69) with the identified research sites (the two academic libraries). This purposive selection included individuals occupying positions such as Senior Librarians (2) of branch library collections (Medicine and Law), Manager (1) of a disciplinary information service

(Commerce), an institutional repository Manager (1), the Head (1) of Digitization Services and a Director (1) of Client Services and e-Scholarship in the university library and information services. The researcher was confident that the occupants of these positions would be a rich source of data needed to respond to the research question generated for this preliminary investigation. Each interview lasted about an hour and was based on a set of questions around disciplinary knowledge, generic skills and personal competencies relating to the interviewee's position as well as the LIS service in general. Depending on the responses from the interviewees, the researcher, where possible, probed further in an attempt to enhance the richness of the data collected.

4. ANALYSIS AND FINDINGS FROM THE PRELIMINARY INVESTIGATION

This preliminary investigation did not focus on job titles and qualifications required but rather scanned the contents of advertisements for requirements relating to disciplinary knowledge, generic skills and personal competencies. These knowledge/skills categories were identified in the literature (Choi and Rasmussen, 2009, Nonthacumjane, 2011 and Orme, 2008) reviewed for this study as being the main categories into which the knowledge and skills requirements for LIS professionals in a digital LIS environment, may be grouped. A chart (see Table 2) was created with these categories and each of these categories was populated with relevant knowledge and skills that emerged in the review of literature for this study. To supplement this and, importantly, for the chart to be a more accurate reflection of the skills requirements for the digital era academic library environment, the ACRL *2012 top ten trends in academic libraries* (ACRL Research Planning & Review Committee, 2012) was consulted and the following three items of disciplinary knowledge were added: scholarly communication, e-resources collection development, and research support librarianship. Further, where necessary, knowledge and skills included in the categories were "iteratively extended and refined" (Wise et al., 2011, p. 275) during the analysis of the data — that is, as the researcher worked through the contents of the job requirements sections of the advertisements and the transcripts of the interviews, where necessary, required skills sets were grouped together under a general attribute if they were part of a generic group of skills or separated into specific attributes if they fell into distinct skills requirements, depending on the pattern that unfolded as the researcher meticulously worked through the content of the job requirements data collected.

The advertisements were scrutinized for requirements and these were allocated to the skills categories on the chart. Data was gathered from the requirements sought by the employers and not from the duties or responsibilities of the post, unless this referred to a particular requirement (Orme, 2008, p. 624). Each requirement was recorded on the tally chart. Once this was done for all the advertisements, the counts were totalled to produce frequency counts allowing for "an assessment of the relative importance of the different requirements and the areas into which they fall" (Orme, 2008, p. 624). A total of 71 advertisements were analysed. A similar process was applied to the data collected from the six interviews (a total of about 6 h of interviewing) except here the researcher had to wade through much qualitative data and had to be mindful of not capturing repetitions. The frequency counts (together with percentages) from both the data sets are reflected inTable 2. Fig. 1, using a bar graph, captures the totals (and corresponding percentages) for each of the skills categories. Juxtaposing the totals from the two data sources (job advertisements and interviews) on the graph allows for comparison of trends between the two data sets, for purposes of triangulation.

4. Analysis and findings from the preliminary investigation

Table 2. Knowledge and skills requirements in the digital era academic library in South Africa.

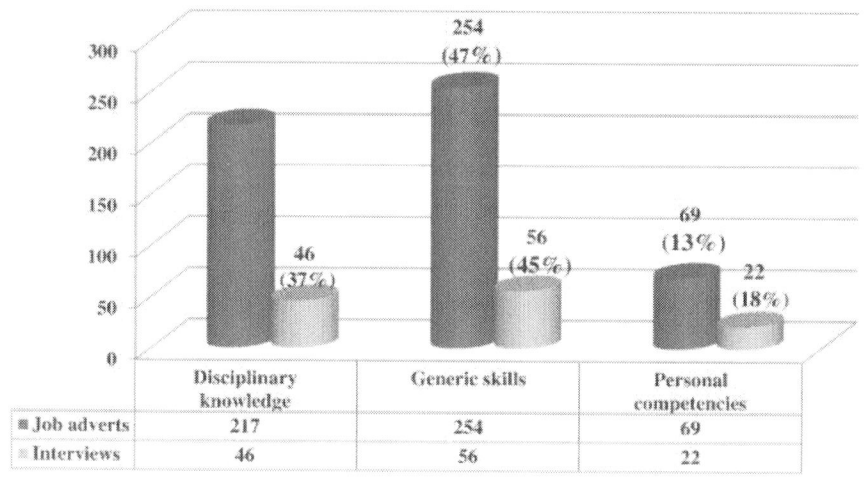

Figure 1. Knowledge and skills requirements in the digital era academic library in South Africa.

5. DISCUSSION, CONCLUSIONS AND LESSONS FOR THE WIDER ACADEMIC LIBRARY SKILLS STUDY

This section discusses the findings from this exploratory study in relation to the literature reviewed and based on this discussion draws conclusions in response to the research question: What key knowledge and skills are required for LIS professionals to effectively and efficiently practise in a digital era academic library in South Africa? A secondary purpose of this preliminary investigation was to tease out some of the parameters for the wider study targeting the development of a comprehensive skills statement for higher education libraries in South Africa. Hence this section also looks at some of the lessons gained from this exploratory investigation for the wider study.

5.1. Discussion and conclusions

As revealed in the literature (Nonthacumjane, 2011, Orme, 2008, Partridge and Hallam, 2004 and Wise et al., 2011) and reflected

5. Discussion, conclusions and lessons for the wider academic library skills study

in Table 1, in this study too it emerges that a variety of skills and competencies are required of the modern LIS professional. A key convergence in findings between this South African study and similar ones in the USA, United Kingdom and Australia cited in the review of literature, is that generic skills emerged as being the most required skills set (see Fig. 1). Table 1, which captures key trends from the literature relating to academic library knowledge and skills requirements, shows a 55% preponderance of the trend of employers increasingly emphasizing generic skills as priority skills requirements. At the same time, Orme (2008, p. 626) observes that despite the 'move to the generic' argument in the literature (Kennan et al., 2006 and Wise et al., 2011), and also evident in the findings of the study reported here, disciplinary knowledge is still valued by LIS employers, as evidenced in their US study by a strong presence of disciplinary skills in the top twenty list of frequently sought skills (see also Table 1 — 45% preponderance of this trend reflected in the literature). The current study lends weight to this argument when one observes in Table 2 the high frequency counts against, for example, 'traditional LIS resources, services and functions' (71 and 21, respectively) in both the job advertisement and interview data sets. This South African study, like those abroad, also saw personal skills, while sought by employers, lagging behind disciplinary knowledge and generic skills in popularity (see Fig. 1). Findings from the literature, summarised in Table 1, reiterate this point by reflecting this as a notable trend in the literature.

'Technology associated with LIS applications in the digital era' as a required knowledge/skills set achieved a high frequency count in the job advertisement data set (see Table 2). This is not surprising given the impact of technology on the academic library of the digital era, as detailed in the literature (Choi and Rasmussen, 2009, Luce, 2008, McCarthy, 2005, O'Connor and Au, 2008, Ocholla and Shongwe, 2013 and Reeves and Hahn, 2010) and reflected in a 64% preponderance of this trend in the literature (see Table 1). While this exploratory study conflated all technology associated with LIS applications into one skills category, in the wider study envisaged it would be useful to separate this into knowledge of and familiarity with systems software on the one hand, and technical skills on the other. Such technical skills would include digital library architecture and software, technical and quality standards, HTML coding, Web mark-up languages such as SGML and XML, and possibly even some skills in programming and scripting languages. Such a separation would be useful to determine to what extent IT technology skills are becoming a necessity in the education and training of academic library LIS professionals. This IT skills set becomes particularly relevant when one considers that the literature is very emphatic about the fact that the digitally oriented academic library is a renewed conceptualisation of traditional LIS resources and services that is now underpinned and driven by new technologies (Choi and Rasmussen, 2009,Gerolimos and Konsta, 2008, McCarthy, 2005, Middleton, 2003 and Missingham, 2006) or that "LIS [particularly in the digital academic library environment] is becoming an increasingly technology-driven profession" (Riley-Huff & Rholes, 2011, p. 129). This would also explain 'Technology associated with LIS applications in the digital era' following closely on the heels of 'Traditional LIS resources, services and functions' in the upper end of frequency counts in the job advertisement data set in Table 2.

The high job advertisement frequency count against 'Experience" in the disciplinary knowledge skill set (42 — see Table 2) implies, as Orme (2008, 629) suggests, that for employers knowledge, skills and personal attributes alone are "insufficient without an ability to demonstrate their practical application". The job advertisement frequency counts for, 'Scholarly communication', 'e-Resources collection development' and 'Research support librarianship' (see Table 2) indicate that these skills areas have taken hold in the modern academic library in South Africa and are on an upward trajectory as suggested in the literature (Choi and Rasmussen, 2009, Luce, 2008 and Nonthacumjane, 2011) about academic libraries in other parts of the world. 'Scholarly communications',

5. Discussion, conclusions and lessons for the wider academic library skills study

particularly, received quite a boost in frequency counts from among 2013 advertisements, indicating an increasing demand in the digital academic library environment for knowledge of new trends in scholarly communications. Combining subject knowledge with professional LIS skills is somewhat of a grey area with just one interviewee in the current study claiming that this is "the ideal" which the LIS service should strive for. Partridge and Hallam (2004) report that "'subject discipline' may becoming more and more important" …especially in fields such as law and medicine where LIS professionals need to "develop or acquire skills and knowledge unique to that field". Not co-incidentally, the frequency count of 10 for subject knowledge requirement in Table 2 emanates largely from law libraries. The rest of the interviewees (five) believed that while having the 'subject discipline' was an advantage, this is "hardly a reality—it is difficult to find a medical graduate working in a medical library".

The emergence on the radar screen of new skills types such as 'Curation' and 'Research data services', albeit with low frequency counts (see Table 2), represent in the South African context too the "emerging trend of research data management and curation" (Wise et al., 2011, p. 279) accompanying the emergence of eScience and eResearch (Luce, 2008), as has been observed by researchers in academic LIS services internationally. South African academic libraries too, like their overseas counterparts are being confronted, in a context of a shortage of skills and training in this area, with the challenge of developing data-archiving infrastructure for the description, management, access and sharing of data (University of Melbourne, 2008). The researcher is convinced that in another year or so, the frequency counts in these new skills areas (digitization, curation, research data service) would rise significantly.

While the generic skills set in Table 2 is led by 'General managerial/supervisory skills' with a frequency count of 48 (in the job advertisement data set), this high figure might be the result of the fact that this skills category, like the 'Traditional LIS resources, services and functions' category in the disciplinary knowledge section, was a generic skills category that pooled together a number of different skills. In the wider study planned, these aggregated categories should perhaps be thinned out for a more accurate reflection of knowledge and skills trends. Notwithstanding this, the generic skill 'Communication' which traditionally is reflected in the literature as the most highly ranked generic skill (Gerolimos and Konsta, 2008, Middleton, 2003 and Orme, 2008), does reflect in Table 2 as a noticeably highly ranked generic skill, both in the job advertisement as well as in the interview data sets. Mention needs to be made of 'Interpersonal skills', 'General computer skills and computer literacy', 'Client service orientation' and 'Teamwork' — these are generic skills which also notched up significant frequency counts (see Table 2) lending weight to the 'move to the generic' trend evident in the literature (see Table 1) as well in the findings reported in this paper (see Fig. 1 and Table 2). This increasing tendency towards a demand for skills relating to generic capabilities, over and above disciplinary or professional knowledge and skills, is also reflected by the longer list of skills requirements in the generic skills set compared to the other two (see Table 2). The 2013 job advertisements, particularly, surfaced noticeable requests for generic skills relating to project management, strategic planning, branding and marketing, budgeting, performance management and evaluation, etc. (see Table 2). Digital academic libraries are expensive enterprises, and hence it is not surprising that these generic skills are in demand to ensure the efficacy of these enterprises.

It would seem that the knowledge and skills required for LIS professionals to effectively and efficiently practise in a digital era academic library in South Africa are a blend of discipline-specific knowledge, generic skills and personal competences, with some of these skills types being more valued in this context than others. At the same time, these LIS professionals, like their counterparts in other parts of the world, are also being challenged by new and emerging skills

5. Discussion, conclusions and lessons for the wider academic library skills study

requirements. Many of the latter are a re-conceptualisation of traditional LIS skills using new technologies and hence it is important for the new generation LIS professionals to use this knowledge base to adapt existing skills to respond to new 'problems' in a working environment that is constantly changing. LIS education and training in South Africa too needs to evolve to meet the challenges of the new knowledge and skills requirements of the digital age academic library in South Africa which is a major employer of LIS graduates.

While the intention of this preliminary study was to ascertain an initial picture of key knowledge and skills sets required for LIS professionals in this environment, a secondary intention was to tease out of this exploratory investigation some of the parameters for the wider study targeting the development of a comprehensive skills statement for higher education libraries in South Africa. Hence this paper ends by highlighting lessons for the wider study.

5.2. Lessons for the wider study

This exploratory study was useful in revealing the limitations of relying on a single data source for job advertisements, no matter how good a source it is. It was also instructive in revealing that in the current digital age, online sources of job advertisements are a must if one wants a "comprehensive picture of the job market" (Reeves & Hahn, 2010, p. 105). Hence the envisaged wider study would need to target multiple sources of job advertisements, such as other newspaper titles, online mailing lists (particularly those of the professional body (LIASA) and relevant websites. This is likely to provide a more complete picture of the LIS employment market in South Africa.

While this small study depended on manually tallying the frequency of skills requirements identified through content analysis, the bigger study (with bigger data sets) could make use of content analysis software such as *Provalis Simstat* and *Wordstat* (Wise et al., 2011, p. 275) used in other successful job advertisement studies. The interviews too were time-consuming, geographically restricting (to the Western Cape) and its more qualitative nature did not blend very naturally with the type of data collected from the job advertisements which lent itself more to quantification and hence more convenient analysis during the triangulation process. Moreover, the interviews, due to limitations to the number that can be conducted over a specified period of time, did not prove, in the end, to be such a productive source of research data for this type of study. Based on this research experience and the knowledge gained from this exploratory exercise, the researcher would suggest that the wider study, instead of gathering data via semi-structured interviews, make use of quantitative survey data collected via the use a structured Web-based questionnaire administered to LIS professional staff from academic libraries scattered across the country, for a truly national picture and a more efficient triangulation of data sources. Content analysis software could be used to analyse the much larger volume of data that is intended for collection. The use of data analysis software with a larger volume of data would also allow for more than just frequency counts. Other more sophisticated analyses such as co-occurrence and multivariate analyses (Wise et al., 2011, p. 283) could be employed, thus allowing for deeper analysis and a more complete picture of the requirements of employers, employment opportunities and emerging trends in the LIS employment market.

REFERENCES

1. ACRL Research Planning and Review Committee (2012). 2012 top ten trends in academic libraries: A review of the trends and issues affecting academic libraries in higher education. Retrieved February 24, 2013, from. http://crln.acrl.org/content/73/6/311.full

REFERENCES

2. Choi, Y., & Rasmussen, E. (2006). What is needed to educate future digital librarians: A study of current practice and staffing patterns in academic and research libraries. D-Lib Magazine, 12(9) (Retrieved May 14, 2012, from http://www.dlib.org/dlib/ september06/ choi/ 09choi.html).

3. Choi, Y., & Rasmussen, E. (2009). What qualifications and skills are important for digital librarian positions in academic libraries? A job advertisement analysis. The Journal of Academic Librarianship, 35(5), 457–467.

4. Gerolimos, M., & Konsta, R. (2008). Librarians' skills and qualifications in a modern information environment. Library Management, 29(8/9), 691–699

5. Kennan, M.A., Cole, F., Willard, P., Wilson, C., & Marion, L. (2006). Changing workplace demands: What job ads tell us. Aslib Proceedings: New Information Perspective, 58(3), 179–196.

6. Luce, R. E. (2008). A new value equation challenge: The emergence of eResearch and roles for research libraries. Retrieved January 22, 2013, from. http://www.clir.org/pubs/ reports/pub142/luce.html

7. McCarthy, J. (2005). Planning a future workforce: An Australian perspective. New Review of Academic Librarianship, 11(1), 41–56.

8. Middleton, M. (2003). Skills expectations of library graduates. New Library World, 104(1184/1185), 42–56.

9. Missingham, R. (2006). Library and information science: Skills for twenty-first century professionals. Library Management, 27(4/5), 257–268.

10. Neuman, W. L. (2006). Social research methods: Qualitative and quantitative approaches. Boston: Pearson.

11. Nonthacumjane, P. (2011). Key skills and competencies of a new generation of LIS professionals. IFLA Journal, 37(4), 280–288.

12. O'Connor, S., & Au, L. -C. (2008). Steering a future through scenarios: Into the academic library of the future. The Journal of Academic Librarianship, 35(1), 57–64

13. Ocholla, D., & Shongwe, M. (2013). An analysis of the library and information science (LIS) job market in South Africa. South African Journal of Libraries and Information Science, 79(1), 35–43, http://dx.doi.org/10.7553/79-1-113.

14. Orme, V. (2008). You will be …: A study of job advertisements to determine employers' requirements for LIS professionals in the UK in 2007. Library Review, 57(8), 619–633.

15. Partridge, H., & Hallam, G. (2004). The double helix: A personal account of the discovery of the structure of [the information professional's] DNA. Paper presented at the ALIA 2004 Biennial Conference, Gold Coast, Australia, 21–24 September 2004 Retrieved January 25, 2013, from http://conferences.alia.org.au/alia2004/pdfs/partridge.h.paper.pdf.

16. Reeves, R. K., & Hahn, T. B. (2010). Job advertisements for recent graduates: Advising, curriculum, and job-seeking implications. Journal of Education for Library and Information Science, 51(2), 103–119.

17. Riley-Huff, D. A., & Rholes, J. M. (2011). Librarians and technology skill acquisition: Issues and perspectives. Information Technology in Libraries, 30(3), 129–140.

18. Tammaro, A.M. (2007). A curriculum for digital librarians: A reflection on the European debate. New Library World, 108, 229–246.

REFERENCES

19. University of Melbourne (2008). Future skills for academic librarians: research and research and training paper for Information Futures Commission. Retrieved March 15, 2012, from. www.informationfutures.unimelb.edu.au/_data/assets/pdf_file/0010/105103/15-research20080509.pdf

20. Welman, C., Kruger, F., & Mitchell, B. (2005). Research methodology (3rd ed.). Cape Town: Oxford University Press.

21. Wise, S., Henninger, M., & Kennan, M.A. (2011). Changing trends in LIS job advertisements. Australian Academic and Research Libraries, 42(4), 268–295.

CHAPTER 8

Library Value in the Classroom: Assessing Student Learning Outcomes from Instruction and Collections

Denise Pan[1], Ignacio J. Ferrer-Vinent[2], Margret Bruehl[3]

[1]Technical Services, University of Colorado Denver, Auraria Library, 1100 Lawrence Street, Denver, CO 80204, USA
[2]University of Colorado Denver, Auraria Library, 1100 Lawrence Street, Denver, CO 80204, USA
[3]University of Colorado Denver, Department of Chemistry, Campus Box 194, P.O. Box 173364, Denver, CO 80217-3364, USA

ABSTRACT

What is the value of library services and resources in the college classroom? How do library instruction and collections contribute to academic teaching and learning outcomes? A chemistry instructor, instruction librarian, and technical services librarian collaborated to answer these questions by combining chemistry education and information literacy pedagogy to assess student learning. The authors developed curriculum units that teach information literacy skills and scientific literature research in a General Chemistry Laboratory course for Honors students. Their study extends beyond examining library instruction and collections assessment in isolation. Rather, their research protocol intends to contribute to student learning outcomes assessment research. The authors propose that an embedded, mixed-methodology, and longitudinal approach can be used to collect data and assess outcomes in terms that describe and measure the value of library services and resources.

KEYWORDS

Value of library services and resources; Assessment; Information literacy; Student learning outcomes; Library instruction; Collections assessment

1. INTRODUCTION

> *The institutional goal of research universities should be a balanced system in which each scholar – faculty member or student – learns in a campus environment that nurtures exploration and creativity on the part of every member* (Boyer Commission on Educating Undergraduates in the Research University, 1998, p. 10).

This story began with a simple email and lead to an interdisciplinary study conducted over three academic years. At the University of Colorado (CU) Denver, in the summer of 2010, a technical services librarian contacted a chemistry instructor and asked: "In the current economic climate, when university administrators are looking for ways to balance the budget, it is imperative that libraries provide evidence of value and demonstrate their contribution to university priorities. Are you available to discuss your potential participation in a study?" By fall term, the survey participant became a co-researcher and with the help of an instruction librarian, the library was embedded into two chemistry classes. Together the three like-minded faculty members from different areas of academia and librarianship collaborated to produce curriculum units for Honors students in General Chemistry Laboratory I and II courses, administered over fall and spring semesters. The authors exposed first-year students to scientific literature and assigned information literacy activities that help build problem-solving and critical thinking skills to engage and promote student success. Moreover, they developed and implemented a research protocol that enabled them to gather and analyze data on student learning outcomes over time.

At the conclusion of a three-year study, the three faculty members asked themselves three questions: Can we generalize this methodology? Will it scale? Does it contribute to the organizational goals of student retention and success? Their answer was, "we believe so." This article is an invitation to practicing librarians and Library and Information Science researchers to implement the CU Denver research protocol for gathering and analyzing data to measure the value of library services and resources. In this article, the authors will explain how their article contributes to the growing body of literature focused on student learning outcomes assessment; describe their research protocol and curriculum units; and provide a summary of study results. Companion articles address the case study methodology, implementation, and student performance assessments (Ferrer-Vinent, Bruehl, Pan, & Jones, submitted for publication); and describe the curriculum units developed and their connection to building information literacy (Bruehl, Pan, & Ferrer-Vinent, submitted for publication).

2. STUDENT LEARNING OUTCOMES ASSESSMENT

Like a rally call to the troops, the Association of College and Research Libraries (ACRL) and Megan Oakleaf created *The value of academic libraries: A comprehensive research review and report* (2010). In essence, the report encourages campus level conversation on assessment, accountability, and value. Within the context of institutional mission and outcomes, they identify "Student Success" as one of the top ten areas of library value on which to focus a research agenda. In response to the question – "How does the library contribute to student learning?" – Oakleaf states that the current literature on information literacy is "voluminous," but a majority is "sporadic, disconnected, and reveals limited snapshots of the impact of academic libraries on learning." Instead, she recommends that "Academic librarians require systematic, coherent, and connected evidence to establish the role of libraries in student learning" (ACRL, 2010, p. 118). In her review and analysis of the literature, Oakleaf introduces several practical suggestions (p. 37–42). The authors distill these concepts into three words that describe the essence of their foray into student learning outcomes assessment: collaboration, purposefulness, and longevity.

3. COLLABORATION

Comprehensive and meaningful assessment of student success is impossible in isolation. A learning ecosystem can be cultivated between student and instructor; student and librarian; and instructor and librarian. Poetically described by the Boyer Commission report, *Reinventing undergraduate education,* "The ecology of the university depends on a deep and abiding understanding that inquiry, investigation, and discovery are the heart of the enterprise, whether in funded research projects or in undergraduate classrooms or graduate apprenticeships. Everyone at a university should be a discoverer, a learner" (Boyer Commission on Educating Undergraduates in the Research University, 1998, p. 9). Throughout library literature, these sentiments have been echoed for more than a decade.

When evaluating the 21st century library, Smith (2001) describes the changing environment of higher education. The concept of learning has shifted from the "teacher's knowledge to the student's understanding and capabilities…it requires the faculty to bring the strength of the research paradigm into the learning process" (2001, p. 29). Academic faculty members are being asked to become learning experts by focusing on outcomes assessment — developing individual students' competencies and demonstrating collective programmatic success. As part of the academic community, the mission of the library must change from "a content view (books, subject knowledge) to a competency view (what students will be able to do)" (p. 32). No longer gatekeepers to materials or tools, academic librarians must take a more active role in the learning process and contribute student learning outcomes for academic programs across the curriculum.

Similarly, Nimon recognizes the expanded role of librarians in measuring the outcomes of academic programs. To do so, she encourages developing partnerships between the library and academic departments to teach information literacy. Moreover, the success is contingent on including assessment criteria that reflect the goals of all stakeholders — librarian, academics, and students. She explains, "Student evaluation of the program must be appropriately tailored to show whether its goals were readily visible to the learners and whether the learners considered them met…It will be necessary for the assessment of student work to be at least in part a joint responsibility" (Nimon, 2001, p. 50).

Participation in pedagogy and assessment activities is not just the role of academic librarians, but their obligation. Bundy asserts that they "can no longer responsibly disengage from why students want the print and digital information and resources to which libraries can now so readily provide access. Nor can they disengage from whether those students have the capacity to apply that information well, and to what use they put it" (2004, p. 2). Instead, both academic teachers and librarians should be immersed in the total educational process — including program and curriculum development, learning design, pedagogies, assessment, and the scholarship of teaching and learning.

Information literacy is an important concept that should be owned by all educators (Bundy, 2004, p. 7). However, this terminology may not be recognizable outside the library walls. Some examples of synonyms and overlapping concepts that may resonate more widely include the following: information, research, or 21st century skills; independent scholarship or research; lifelong learning; scientific method; research processes; and Bloom's taxonomy. As Oakleaf explains, "For those facing greater challenges, establishing and using a common language that emphasizes shared campuswide values may produce greater success" (2011, p. 65).

In his model for academic libraries, Lewis (2007) proposes a strategy for maintaining the central library position on campus in the digital age. One of his top five strategies is to "Reposition library and information tools, resources, and expertise so it is embedded into the teaching, learning, and research enterprise...Emphasis should be placed on external, not library-centered, structures and systems" (2007, p. 3). By focusing on student learning, academic libraries and librarians have new opportunities to reestablish their place on campus, engage with their colleagues, and maximize their contribution to their institutions and higher education as a whole.

4. PURPOSEFULNESS

When developing student learning outcomes assessments, academic librarians, and faculty members should proceed with purpose. In other words, they are intentionally gathering and analyzing particular types of data. Specifically, Dugan & Hernon (2002)assert, "student learning outcomes are concerned with attributes and abilities, both cognitive and affective, which reflect how the student experiences at the institution supported their development as individuals" (2002, p. 377). In addition, "outcomes assessment alerts us to what students know or do not know about research," Carter explains, "thus allowing librarians to adapt instruction to the needs of the students" (2002, p. 41). This requires a commitment to document, evaluate, and communicate impacts on student learning, as well as improving one's own teaching and assessment skills (Oakleaf, 2011, p. 70).

The literature on information literacy assessment identifies numerous techniques and tools. These tactics require a range of resources or experiences and assess various perspectives. Radcliff et al. describe three different learning domains: *affective* assesses how students feel or their opinions; *behavioral* evaluates what students can do; and *cognitive* measures what students know. They classify performance assessments, such as report writing, as part of the behavioral domain (2007, p. 19–20, 115). In contrast Oakleaf describes performance assessments as "real-life applications of knowledge and skills" and "[they] reinforce the concept that what students learn in class should be usable outside the classroom" (2008, p. 239). All can agree that assignments with an information literacy component can be used to measure higher-order thinking skills, and to achieve greater integration and contextualization in an academic course. While the results can offer a high degree of validity, they may have limited generalizability (Radcliff et al., 2007 and Oakleaf, 2008).

The literature on assessment of information literacy is quite bountiful — well documented in monographs, handbooks, manuals, guides, and articles. The focus is nearly exclusively on library instruction, and overlooks other areas of librarianship. Library collections are included in the broader context of library assessment and emphasize cost effectiveness (Hufford, 2013, p. 20–26). Kinman's article "E-metrics and library assessment in action" is a rare example that highlights the significant role electronic resources play in demonstrating the value of libraries and impacting on student learning outcomes (2009). There is ample opportunity to expand this area of the scholarship into a new line of inquiry that captures the complexity of the learning environment and inspires more rigorous and critical investigation.

5. LONGEVITY

More than library instruction, information literacy extends outside the confines of the library and extends to the classroom, campus, and beyond. Each level requires more collaboration and complex logistics (Iannuzzi, 1999, p. 304). To be truly meaningful, outcomes assessment is an ongoing process. It requires time, effort, resources, and collaboration with academic faculty. The research process can be simple or elaborate, but must be based on meaningful data. From each experience, researchers can learn, refine, and improve assessment procedures (Carter, 2002, p. 41). Oakleaf & Kaske (2009) recommend that librarians follow best practices, use multiple methods and strategies, adjust goals and objectives, and repeat assessments continuously over time (2009, p. 283).

6. CU DENVER RESEARCH PROTOCOL & CURRICULUM UNITS

The literature describes collaboration, purposefulness, and longevity as key ingredients to achieve student learning outcomes assessment. This article continues this conversation. The authors at CU Denver adopted these assessment perspectives and adapted them to their local circumstances. Specifically the authors summarize their student learning outcomes assessment research protocol in three components: collaborating internally and externally; articulating outcomes; and assessing over time. By doing so they are able to expand beyond their silos and approach student learning outcomes assessment from an interdisciplinary perspective. The results of their study demonstrate the rich data and insights that are possible from their approach.

7. COLLABORATING INTERNALLY AND EXTERNALLY

The CU Denver researchers identified multiple opportunities for collaborations within librarianship, academia, and the campus. They found commonality with academic faculty and librarians from diverse areas and expertise. In addition, they also created teaching, research, and learning opportunities between faculty and students. A visual representation of these connections is available in Fig. 1.

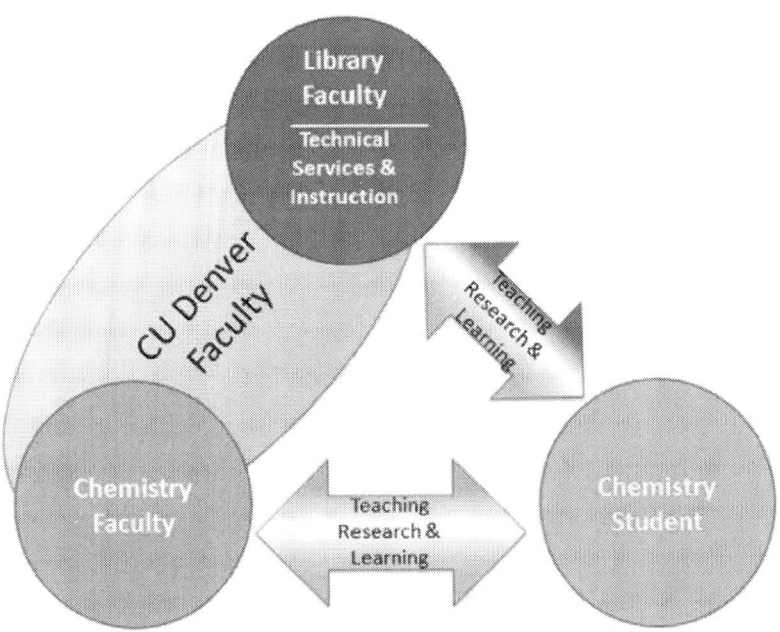

Figure 1. Research relationships.

At CU Denver, librarians have faculty status. The library and chemistry departments are two separately administered academic units reporting to the Office of the Provost. Although the departments often cooperate on collections and instruction initiatives, the authors' research study represents a new level of collaboration. Library instruction and collections assessment activities are embedded into two Chemistry courses. More specifically, the researchers focused on shared values and appreciated other perspectives from different disciplines. In essence, they are proponents of a "golden triangle" approach to curriculum design "where everyone contributed to each component with the outcome being more robust than if they used the more linear model" (Fox & Doherty, 2012, p. 151–2).

Within the library context, a technical services librarian and an instruction librarian from the same academic library developed a partnership to conduct a student learning outcomes assessment research study with a chemistry instructor. This experience may be atypical. Bundy states that "few university libraries or librarians directly engage with, or reach out to, other parts of the profession…the focus tends to be on information resource sharing and access, rather than on learning collaborations and strategies" (Bundy, 2004, p. 9).

Most significantly, the CU Denver faculty developed a research relationship with their students. With Colorado Multiple Institutional Review Board approvals, they were granted permission to gather data from the students. In her Post Laboratory Assignment, Bruehl specifically invited her students to participate in the study and assured them that no additional work would be assigned, "Simply completing this assignment is all that is required." All personal identification was removed by Bruehl before Pan and Ferrer-Vinent analyzed the student data. Bruehl also explicitly told the students that they could contact her during lab or via email to option out of the study and their data would be excluded from the study (Bruehl et al., submitted for publication).

8. ARTICULATING OUTCOMES

Sharing common values and goals, the authors found exciting synergies for developing student learning outcomes assessment. All three researchers aspired to develop a project that could provide compelling evidence of supporting the institution's mission for campus administrators. They surmised that an assignment that concentrated on information literacy could help them achieve this objective. In doing so, they developed a research instrument that enabled them to gather data on students' ability to find, access, and evaluate scientific literature.

When developing her Honors General Chemistry I and II Laboratory courses, Bruehl aims to create a student centered learning environment. This teaching philosophy encourages active participation by students while the learning process is facilitated by the instructor. The investigative and inquiry-based laboratory courses emphasize teaching students problem-solving and critical thinking skills through open-ended experiments and using specialized techniques and instrumentation. This teaching and learning strategy builds on the CREATE (Consider, Read, Elucidate hypothesis, Analyze and interpret data, Think of the next Experiment) method.

Typically, the real language and research process documented in primary scientific literature is not presented to students until their upper division course or not at all during their undergraduate education. This delay is recognized as a missed opportunity at best, or a loss of science majors at worst. Using the CREATE pedagogy exposes students to scientific research literature and demonstrates the creative and exploratory nature of collecting and interpreting data for research. By doing so, students are being taught how to think like scientists, increase engagement, and could help with retention (Hoskins et al., 2007 and Gottesman and Hoskins, 2013). Findings from one study indicate that the CREATE method "increases students' confidence in their ability to read and understand primary literature, improves their self-assessed understanding of the nature and processes of science, and encourages their development of more sophisticated epistemological beliefs" (Hoskins, Lopatto, & Stevens, 2011, p. 375).

The CU Denver research study has two overarching goals. The first is to demonstrate that teaching information literacy skills in first-year chemistry courses can provide immediate and long-term benefits to student performance. The second is to quantify the benefits students received in their educational activities by using library collections. Long term, the authors hope to be able to provide campus administrators with evidence that their efforts contribute to meaningful outcomes, such as student retention, graduation rates, and economic benefit.

To achieve these goals, the authors developed two complimentary curriculum modules to establish a foundation of information literacy skills for beginning science students using scientific literature and laboratory experiments. The first unit, administered in the fall semester, is entitled "*Introduction to Scientific Literature*" and is comprised of three components: 1) formal library instruction by a science librarian; 2) reading and in-class discussion on a scientific journal article; and 3) pre- and post-lab exercises to research an idea for a new general chemistry experiment. The students are asked to develop information literacy skills by searching the scientific literature for concepts, recording their research process, refining their premise, writing a brief description of their proposed experiment, and citing three resources in American Chemical Society (ACS) format that directly support their proposal.

Building on the skills and information they learned in the fall, in the spring semester they are asked to conduct a multi-week project entitled "*Design your own General Chemistry Lab.*" Again students explore the scientific literature for ideas on which to base their experiment, and record the reference for any resources they downloaded and read. Their final project includes a formal proposal; documented laboratory procedure that has been designed, developed, and tested by the student; citation list in ACS format; and presentation of results to the class.

In both semesters, students are conducting research in the scientific literature. To record their research process, Bruehl requires students to complete a research process template. The format of the template is a table with three separate columns to record: 1) database/search tool name; 2) search terms/refinements; and 3) resources viewed. The third column instructs the students to "paste ACS or specific journal format citation here."Table 1 includes an example of a completed research process template, with good searching and narrowing values, provided by a student in spring 2012.

The research process template is an essential assessment tool for the authors to gather data to evaluate the students' ability to find, access, and evaluate scientific literature. Bruehl required the student to record *all* resources viewed, even if the student only read the abstract or the table of contents, and not the article itself. She explains that, "Some resources that you record in this template will be more important to your project. These resources help to formulate your design, and they must ALSO be listed in your formal citation list" (Bruehl et al., submitted for publication).

By documenting the names of database/search tool, search terms, and resources viewed, the authors could see how students are selecting the resources viewed — rejecting or accepting results that fall in and out of the scope, or distract or meet the needs of their topic. Broad subjects are narrowed with keywords, and then filtered down by publications focused on science experiments, such as the *Journal of Chemical Education*. Combining the expertise from academic faculty and faculty librarians, the authors analyze and compare the data from the research process templates and citation lists. Bundy summarizes this partnership and librarian contributions to teaching pedagogy as "Disaggregated roles, such as assessing learning resources for quality, overlap with what librarians do now, and the subject expertise of the academic teacher is being married with the librarian's navigation and sense making of the information universe" (Bundy, 2004, p. 9).

Table 1. Sample research process template.

9. ASSESSING OVER TIME

Realizing that outcomes assessment is a continuous process, the CU Denver researchers intentionally developed the project to be repeatable over multiple semesters and academic years. The process encourages collaborative teaching and learning from the students and one another. The authors refine and improve instructional strategies and assessment methods at each encounter. For example, during citation analysis Ferrer-Vinent and Pan observed that a few students were citing articles that were not available full-text from the library. While it is possible that the students could have requested the article from interlibrary loan, the librarians were worried that these students did not understand the difference between an abstract and the full-text article. They shared their concerns with Bruehl who added this information to her lecture on scientific literature.

Exposing students to information literacy concepts over multiple semesters reinforces skills and the interconnectedness of information. The importance of the research skills and process is reiterated over two terms of an academic year. In the spring semester *"Design your own General Chemistry Lab"* assignment, Bruehl reminds students to expand their understanding beyond their own experience, and to use the research expertise they developed the previous semester. Scientific literature offers "a vast array of resources" to assist them with their assignment, which in turn helps them to develop and recognize sound scientific procedures, evaluate experimental methods, and draw appropriate conclusions. As a result, beginning chemistry students are fine tuning their ability to use the scientific method to investigate a scientific question.

Moreover the CU Denver study was conducted over three academic years. The researchers analyzed consistencies and variations over the same and consecutive semesters across multiple academic years. For example they could examine data for each term (e.g. Fall 2010 and Spring

2011), or comparison between terms (e.g. Fall 2010 versus Spring 2011), and within an academic year (e.g. 2010–2011). Most importantly, since the chemistry students agreed to participate in the research project, the researchers surveyed the same students 1–2 years after they finished the Honors General Chemistry II Laboratory course. Therefore, with the CU Denver research protocol, the researchers gathered quantitative and qualitative data from the students at different points of their academic careers — during their first-year and beyond.

10. CU DENVER RESEARCH STUDY RESULTS

On their own, the researchers could have independently assessed outcomes from their own disciplines' perspectives. By collaborating internally and externally, articulating outcomes, and assessing over time, the authors developed and implemented their research protocol from fall 2010 to spring 2013. Expanding beyond their silos and using an interdisciplinary perspective and mixed methodology approach enabled them to produce distinct student learning outcomes assessment data and analysis.

Prior to their partnership, each author individually engaged in assessment activities. Bruehl conducted student learning outcomes assessment by grading students' assignments, exams, and overall performance in the class. Ferrer-Vinent provided in-class library instruction, surveyed student perceptions, and assessed students' post-instruction performance. With no direct contact with students, Pan measured the library's potential contributions by analyzing usage statistics of electronic resources.

Together they leveraged personal expertise to enhance their shared research project. Their interdisciplinary approach allowed them to develop and evaluate richer data sets. The CU Denver researchers apply the quantitative citation analysis data from the research process template and qualitative surveys to address their research inquiries. Specifically, they intend to demonstrate that teaching information literacy skills in first-year chemistry courses can provide immediate and long-term benefits to student performance; and to quantify the benefits students received in their educational activities by using library collections. Research outcomes are summarized in three subsequent sections: 1) quantitative information literacy study results; 2) quantitative collections ROI study results; and 3) qualitative student benefits survey results. As previously mentioned, details on the authors' research methodology is described in their companion case study article (Ferrer-Vinent et al., submitted for publication).

11. QUANTITATIVE INFORMATION LITERACY STUDY RESULTS

By recording the references of resources viewed and cited in their assignment, the authors could begin to quantify and aggregate how students are developing information literacy skills. For example, Table 2 demonstrates that they were able to calculate the average number of resources viewed per student to complete the fall and spring term assignments and their average course grade. For the fall semester assignment, students were instructed to use their library skills to view resources and then select three upon which to base a proposed experiment. In spring, students received no guidance on how many resources should be viewed or selected to support their experimental design. Excluding the first semester, during which the authors were still refining the assignment guidelines, students on average looked at almost 10 resources in the course of their assignments and their average grade was 87.5%.

Table 2. Resources viewed and final course grade.

The researchers intentionally collected grades to enable them to correlate academic performance and use of library resources. Said another way, they hypothesize that there is a relationship between the extensive use of library resources and a high score. If students consulted multiple databases, reviewed many articles, and selected the best three resources, they should receive a higher score than a student who only used one database and looked at a few articles. The authors collaborated with Galin Jones – Associate Professor at the University of Minnesota, School of Statistics – to determine that there is a statistically significant positive relationship between resources viewed and the students' final grade for the course. The full explanation on the linear regression analysis is available in the case study companion article (Ferrer-Vinent et al., submitted for publication).

12. QUANTITATIVE COLLECTIONS ROI STUDY RESULTS

The authors also use the citation analysis data to calculate the cost benefits of purchasing scientific literature to support learning objectives. To do so, they measure library value by deriving value using return on investment (ROI). According to Tenopir (2012), "Derived values, such as return on investment (ROI), use multiple types of data collected on both the returns (benefits) and the library and user costs (investment) to explain value in monetary terms" (2012, p. 6). In order to measure return on investment, the authors needed to assign a monetary value to having access to non-market resources/services or library collections. They applied contingent valuation as defined by Megan Oakleaf to assign value to library collections and to determine potential willingness to pay to maintain the existence of library collections (ACRL, 2010, p. 50). Many publishers, including ACS, sell individual articles online. If the library did not provide these articles, students could purchase these ACS articles on their own. The authors utilize the price to download an article to derive a market value for the library service of providing access to collections. In turn, this calculated value enabled the researchers to calculate student benefits for the ROI calculations.

The CU Denver Student ROI Model is based on established ROI and cost benefit analysis (CBA) formulas. ROI is calculated as a percentage. It shows the return or increase in value on dollars spent to achieve a benefit. The generic formula is benefits minus costs divided by costs and multiplied by 100.

$$((\text{BENEFITS}-\text{COSTS}) \div \text{COSTS}) \times 100 = \text{ROI}.$$

CBA uses the same values as ROI. However, CBA is the ratio showing the dollar value of benefits gained for dollar value of costs. The basic formula is benefits divided by costs.

$$\text{Benefits} \div \text{Costs} = \text{CBA}.$$

Benefits are the estimated cost for students to buy cited articles directly from the publisher with pay-per-view. Costs are the Library costs to supply cited online journals. This methodology is based on a parallel CU Faculty ROI Model and multi-campus study Pan conducted to measure the institutional value of library resources used by faculty in their research (Pan et al., 2013).

The authors analyzed the citations and determined that students viewed 1489 resources over three years. Most of these articles were searched on the ACS Journal Database 58.0% of the time (863 out of 1489). Collectively the students looked at 1153 journal articles from 200 unique journal titles, and they consulted the *Journal of Chemical Education* (*JCE*) 66.7% of the time (769 out of 1153). Since a majority of the students used the *JCE*, the use and cost estimates for this journal title was used to calculate the ROI and CBA for the course. The authors derived the student costs from the number of times *JCE* was viewed, multiplied by the pay-per-viewed cost. The library cost for *JCE* was estimated by taking the database cost divided by the number of titles in the database.

Since ROI is calculated as a percentage, values over 100% are a positive ROI. Similarly, since CBA is a ratio showing how much is gained for every dollar spent, $1.00 is the breaking point. Values greater than $1.00 are a "positive" CBA. The results of the CU Student ROI Model for Honors General Chemistry I and II Laboratory courses at CU Denver from fall 2010 to spring 2013 are summarized in Table 3. Overall, the results indicate a very strong ROI and CBA for the *Journal of Chemical Education* with the greatest ROI of 2324% and CBA of $24.24 in fall 2011.

Table 3. Students ROI study results for *Journal of Chemical Education*.

	ROI	CBA
AY 2011		
Fall 2010	626%	$7.26
Spring 2011	884%	$9.84
AY 2012		
Fall 2011	2324%	$24.24
Spring 2012	1109%	$12.09
AY 2013		
Fall 2012	1971%	$20.71
Spring 2013	288%	$3.88[a]
Total	1144%	$12.44

[a] The researchers attribute the low ROI of 288% and CBA of $3.88 in spring 2013 to low student enrollment due to a scheduling conflict and cancellation of one section of the course.

13. QUALITATIVE STUDENT BENEFITS SURVEY RESULTS

In addition to collaborating on quantitative citation and cost benefit analysis, the researchers also conducted qualitative assessments. After the students completed the Honors General Chemistry II Laboratory course, Bruehl emailed them an anonymous survey. This was an opportunity to gather data on the impact of teaching information literacy to first-year chemistry students over the long term. The questionnaire reminded the students that "as part of your lab work, you used the [scientific] literature and captured your on-line searches in a research process template," and invited them to participate in the survey as a "follow up to those activities." A copy of the longitudinal student survey is available in the companion article focused on the curriculum units (Bruehl et al., submitted for publication).

The responses to the survey were overwhelmingly positive. Over 40% (37 out of 88) of the students completed the questionnaire. When asked about their experiences before and after Honors General Chemistry Laboratory, more than half stated that they searched a scientific literature database since course completion. In addition, this was nearly a 40% increase over their experience prior to taking the course (see Table 4).

Surprisingly, students reported that they searched the American Chemical Society (ACS) journals for scientific literature more frequently than Google (77.4% and 71.0% consecutively). SciFinder, ScienceDirect, and Web of Science were also identified as scientific literature databases that they used after finishing the course. The two other resources identified were PubMed and JSTOR (more details provided in Table 5).

More or less than half of the students responded that they did the search for their own curiosity or for a research project/internship (see Table 6). The vast majority of students, however, conducted these database searches for another course. The knowledge they acquired was directly applied to 17 other Chemistry courses. More significantly, their information literacy skills transferred to 21 courses in Biological Sciences, 2 in English, and 1 in Psychology. These findings support the authors' hypothesis that teaching information literacy skills can provide students with long-term benefits.

Table 4. Reponses to follow up survey.

Table 5. Database usage since completing Honors General Chemistry Laboratory.

Table 6. Explanation for database searches.

	# of responses that searched database *since* class	Explanation			
		Course	Research project/internship	Curiosity	Other
AY 2011	8	8	3	6	
AY 2012	17	13	9	9	1
AY 2013	6	7	2	1	
Total #	31	28	14	16	1
% of Total		90%	45%	52%	3%

14. CONCLUSION

With the CU Denver student learning outcomes assessment research protocol, three researchers from different areas within librarianship, academia, and the campus collaborated to explore the value of library services and resources in the college classroom. Moreover, a chemistry instructor, instruction librarian, and technical services librarian sought to create a methodology for describing and measuring the benefits of library instruction and collections to academic teaching and learning. They discovered that their process includes three components – collaborating internally and externally, articulating outcomes, and assessing over time; and echoes the elements identified in the student learning outcomes assessment literature – collaboration, purposefulness, and longevity.

This study demonstrate the possibilities when academic faculty and librarians move beyond established roles and responsibilities, and attempt student learning outcomes assessment from an interdisciplinary perspective. The authors developed a research process template to gather data used to evaluate the students' ability to find, access, and evaluate scientific literature. In turn, this citation analysis was used to assess the efficacy of teaching information literacy skills in first-year chemistry courses and the value of purchasing scientific literature to support learning objectives. With their mixed methodology approach, the research also combined qualitative assessments with their quantitative citation and cost benefit analysis.

The CU Denver research protocol has just begun to scratch the surface by focusing on assessing information literacy outcomes within the library and the chemistry laboratory. What would happen if similar curriculum units were adapted and applied to different disciplines? Can we teach students to think like historians, sociologists, economists, etc. by exposing them to the research literature? There is ample opportunity to extend this protocol to other disciplines and beyond their institution. Iannuzzi explains that, "…if we want to ensure that those skills are applied within other courses, that there is meaningful transfer to other learning environments, and that ultimately the quality of the student's work is improved, the assessment methodology moves beyond library control into collaborative efforts with teaching faculty" (Iannuzzi, 1999, p. 304). The CU Denver experience is an example of engaging with faculty, asking for input, suggesting a partnership, and co-creating an interdisciplinary study. In essence, exciting opportunities can be possible when librarians leave the library, meet and engage with academic faculty, and capitalize on serendipity.

ACKNOWLEDGEMENT

We would like to thank the University of Colorado Denver Honors General Chemistry I and II Laboratory students who provided the data for this article.

REFERENCES

1. Association of College and Research Libraries (ACRL) (2010). Value of academic libraries: A comprehensive research review and report. Researched by Megan Oakleaf. Chicago: Association of College and Research Libraries (Retrieved from www.acrl.ala.org/value).

References

2. Boyer Commission on Educating Undergraduates in the Research University (1998). Reinventing undergraduate education: A blueprint for America's research universities. NY: Stoney Brook (Retrieved from ERIC database. (ED424840)).

3. Bruehl, M., Pan, D., & Ferrer-Vinent, I. J. (2014). Demystifying the Chemistry Literature: Building Information Literacy in first-year Chemistry Students through StudentCentered Learning and Experimental Design. Journal of Chemical Education (submitted for publication).

4. Bundy, A. (2004). Beyond information: The academic library as educational change agent. Paper presented at the 7th International Bielefeld Conference ((February 3–5)Germany. Retrieved from http://conference.ub.uni-bielefeld.de/2004/proceedings/bundy_rev.pdf).

5. Carter, E. W. (2002). "Doing the best you can with what you have:" Lessons learned from outcomes assessment. The Journal of Academic Librarianship, 28(1), 36–41.

6. Dugan, R. E., & Hernon, P. (2002). Outcomes assessment: Not synonymous with inputs and outputs. The Journal of Academic Librarianship, 28(6), 376–380.

7. Ferrer-Vinent, I. J., Bruehl, M., Pan, D., & Jones, G. (2014). Introducing Scientific Literature to Honors General Chemistry Students: Teaching Information Literacy and the Nature of Research to First-Year Chemistry Students. Manuscript submitted for publication in the Journal of Chemical Education.

8. Fox, B. E., & Doherty, J. J. (2012). Design to learn, learn to design: Using backward design for information literacy instruction. Communications in Information Literacy, 5(2), 144–155.

9. Gottesman, A. J., & Hoskins, S. G. (2013). CREATE cornerstone: Introduction to scientific thinking, a new course for STEM-interested freshmen, demystifies scientific thinking through analysis of scientific literature. CBE Life Sciences Education, 12(1), 59–72.

10. Hoskins, S. G., Lopatto, D., & Stevens, L. M. (2011). The C.R.E.A.T.E. approach to primary literature shifts undergraduates' self-assessed ability to read and analyze journal articles, attitudes about science, and epistemological beliefs. CBE Life Sciences Education, 10(4), 368–378.

11. Hoskins, S. G., Stevens, L. M., & Nehm, R. H. (2007). Selective use of the primary literature transforms the classroom into a virtual laboratory. Genetics, 176(3), 1381–1389.

12. Hufford, J. R. (2013). A review of the literature on assessment in academic and research libraries, 2005 to August 2011. Portal: Libraries and the Academy, 13(1), 5–35.

13. Iannuzzi, P. (1999). We are teaching, but are they learning: Accountability, productivity, and assessment. The Journal of Academic Librarianship, 25(4), 304–305.

14. Kinman, V. (2009). E-metrics and library assessment in action. Journal of Electronic Resources Librarianship, 21(1), 15–36.

15. Lewis, D. W. (2007). A model for academic libraries 2005 to 2025. Paper presented at "Visions of Change" ((January 26, 2007), Sacramento, California. Retrieved from http://hdl.handle.net/1805/665).

16. Nimon, M. (2001). The role of academic libraries in the development of the information literate student: The interface between librarian, academic and other stakeholders. Australian Academic & Research Libraries, 32(1), 43–52.

17. Oakleaf, M. (2008). Dangers and opportunities: A conceptual map of information literacy assessment approaches. portal: Libraries and the Academy, 8(3), 233–253.

18. Oakleaf, M. (2011). Are they learning? Are we? Learning outcomes and the academic library. The Library Quarterly, 81(1), 61–82.

19. Oakleaf, M., & Kaske, N. (2009). Guiding questions for assessing information literacy in higher education. Portal: Libraries and the Academy, 9(2), 273–286.

20. Pan, D., Wiersma, G., Williams, L., & Fong, Y. (2013). More than a Number: Unexpected Benefits of ROI Analysis for Collection Development. The Journal of Academic Librarianship, 39(6), 566–572.

21. Radcliff, C. J., Jensen, M. L., Salem, J. A., Burhanna, K. J., & Gedeon, J. A. (2007). A practical guide to information literacy assessment for academic librarians. Westport, CT: Libraries Unlimited.

22. Smith, K. R. (2001). New roles and responsibilities for the university library: Advancing student learning through outcomes assessment. Journal of Library Administration, 35(4), 29–36.

23. Tenopir, C. (2012). Beyond usage: Measuring library outcomes and value. Library Management, 33(1/2), 5–13.

CHAPTER 9

University Libraries and the Development of Lecturers' and Students' Information Competencies

José Antonio Gómez Hernández

Lecturer in Library Science, University of Murcia

ABSTRACT

This article explains why university libraries assume, as one of their priorities, the development of lecturers' and students' information competencies. It also explains some of the options for achieving that goal.

KEYWORDS

information competencies, university libraries, university lecturers, university students, library services, information literacy

1. INTRODUCTION

Today, Spanish university libraries believe that their function goes beyond the mere provision of instrumental support for teaching and learning activities. Theirs is an intrinsically educational function, in which they get wholly involved. Indeed, they have made it one of their priorities (strategic plan, REBIUN, 2007). Libraries are a resource for documentary research, and they should teach people how to do it properly. It is not simply about how to find information, but also about how to evaluate, select, rework, use and communicate it. In other words, it is a matter of contributing to information competency training, which includes procedural, conceptual and ethical aspects alike. It is fundamental to active, constructive and situated learning. Libraries are making a considerable effort to achieve this goal: they have digital repositories for both research and learning, they have implemented a resources centre model for learning and research, they carry out more and more general and specialised user training, they publish tutorials and guidelines for information management, they organise thematic digital resources, they provide OpenCourseWare, they promote reference services, they use social networks, they provide support for lecturers to prepare new teaching materials and so on. This, therefore, is the broad, evolving process covered in this article.

2. WHY DO LIBRARIANS WANT TO ASSUME EDUCATIONAL DUTIES FOR INFORMATION COMPETENCIES?

This new role may, in the first instance, seem rather strange to the university community, because one of the characteristic features of our higher education system is the considerable compartmentalisation of know-how and functions. The fact that librarians want to "teach" as "learning mediators" represents:

- For some lecturers, the potential for their duties to be usurped, because the more formal, regulated teaching function corresponds to them: establishing what a student ought to learn, what content

2. Why Do Librarians Want to Assume Educational Duties for Information Competencies?

should be covered, how to convey it and how to evaluate attainment. Many lecturers still consider libraries to be nothing more than a container of organised resources at the service of their duties and, through their mediation, the tasks assigned to their students.

- For students, something unexpected, because libraries have traditionally been spaces for collective study, with a number of infrastructure facilities for developing habits like working with others, getting hold of recommended materials and so on. However, they have never seen librarians as teachers in the strictest of senses. At the most, they may have considered librarians as assistants or advisors to help them find information when they had special or specific requests.

- For librarians themselves, an effort to adapt to the changes, because their professional development and self-concept has transformed them into organisers, processors, technicians, intermediaries and so on.

Consequently, becoming teachers is, for librarians, a shift of profession and model, and one that is both significant and difficult, like any other. It is a matter of becoming involved in duties carried out by others – albeit not explicitly requested – in a competitive context, since everything has a financial and organisational impact (What recognition is there? What evaluative legitimacy is there? What other departments and services are they working for or against?).

Thus, we come back to the why and the wherefore. There are conceptual and practical reasons. The former are connected with the evolution of a library's mission in line with changes taking place in the university context in which it is located. The latter are connected with the need for a library to justify its existence, to demonstrate that it is worthy of the investment made in it, and to gain greater protagonism than other university services. It is a matter of engaging in a commitment to universities and of becoming relevant.

The world of education and the world of information have changed tremendously over the last 40 years. Means of accessing and consuming information have evolved to such an extent that libraries are forced to restructure their function so as not to become obsolete. Accessible information networks and the spread of hypertext reading, with a host of new ways to interrelate and incorporate knowledge; the ease of communicating and publishing information, opinions and knowledge of differing value without any filters or intermediaries; the demands of the productive system and the labour market with respect to graduates' competencies; the excessive and fragmented nature of information; the availability of virtual campuses where lecturers and students share learning materials and so on. Together, what impact does all of this have on libraries?

- Changes to intermediation duties. Services of a merely intermediary nature are brought into question, given that people can and want to get direct, immediate access to knowledge. A user may not need a library to obtain information. Universities' virtual campuses on the one hand, and the collections of resources that lecturers make directly available to students via subject websites on the other, mean that students do not go to libraries to look for those materials. This leads to a decrease in traditional lending. Likewise, the existence of a large quantity of free information sources and documents in open access archives on the Internet also means that students and lecturers make less physical use of libraries. Moreover, they do not seem to take library web portals as a basic point of reference for obtaining the information they need. On the contrary, they consider, albeit mistakenly, that they have everything on the Internet and in search engines, and that they do not need library web portals.

- This compels libraries to offer innovative, addedvalue services:

- The provision of new working and learning spaces: despite the intangible, virtual nature of information, people want places where

2. Why Do Librarians Want to Assume Educational Duties for Information Competencies?

they can interact, talk and exchange ideas with their lecturers, and where they can get support and technical, methodological and educational advice for the creation of knowledge. Libraries provide all of this through Learning and Research Resources Centres (CRAI).

- The edition and publication of an institution's digital content, with open access. Libraries promote digitalisation and publication of digital content via digital repositories. These repositories broaden the distribution of theses, journals, conference proceedings and other documents published by members of their university communities through initiatives that foster self-archiving, open access and the use of protocols for optimum document gathering. This promotes an ethic of disseminating scientific knowledge created with public funds.

- The selection and filtering of high-quality content: this role is becoming more and more crucial to users; if a library manages to do this, it becomes an accredited source of relevant, accredited content.

- Cultural facilitation. Traditional cultural activities run by libraries are on the up once again: reading clubs; creative writing workshops; painting, photographic, scientific dissemination and social awareness exhibitions; book, music and film collections; literary or other artform competitions and so on are ways of attracting people to libraries and of making libraries a useful social space. This enriches the function of spaces and services, over and above their curricular learning use.

- Participation in the attainment of basic and generic competencies. The above-mentioned services benefit the students' integral education, their personal maturity, the development of creative facets, critical thought and citizenship habits and values, cultural practices, and social and disciplinary interrelations.

- Teaching a competency that is specifically linked to library services, "information management and use", the mastery of which is considered to be an essential attribute for any university graduate in the framework of educational models arising from the European Higher Education Area (EHEA).

The conclusion is that libraries are interested in information competencies because today, rather than physical and digital collections in real or virtual spaces, libraries are places where a group of professionals aspire to ensure that students learn, enabling them to become competent in digital and information skills while they are at university and throughout their lives.

3. BUT HOW DO STUDENTS AND LECTURERS ACCESS AND USE INFORMATION?

An object of study and a question that interests librarians a great deal is the information behaviour of its users. If they understand it, they can identify users' expectations and needs and habits. This will allow librarians to adapt to users in order to guide and improve their practices through information literacy services. These services have been around for over 30 years in Spanish documentary institutions (Pinto, Cordón & Gómez, 2010). Although there are many traits particular to the behaviour of information searching and use (depending on the knowledge area, course, degree or activity of study or research), from works such as those cited in the bibliography at the end of this article, it is possible to obtain a general picture of the identifying traits most common to university lecturers and students in connection with our topic.

3.1. Students

University students are familiar with digital reading based on hypertext browsing; in addition to reading, they create, publish and share content

3. But How Do Students and Lecturers Access and Use Information?

by taking part in networks; they like immediate access, anytime, anywhere, through simple interfaces without intermediaries, through search engines rather than library portals; they are able to multitask, though they skim the information, spending more time on browsing than on reading the information displayed; they usually download and save information that they do not have time to read later; their speed of communicating and viewing information is greater than their in-depth critical capacity. Specifically, the CIBER report (British Library & JISC, 2008), OCLC (2006) and University of Seville (2009) identified a number of shortcomings among new students:

- A poor understanding of information needs and, therefore, difficulties in developing effective search strategies.

- A lack of reflection on the problem to be solved and what its application is going to be leads to impulsiveness in superficial searches using natural language rather than keywords, with a loss of relevant information. R5 A lack of evaluation of the suitability, accuracy, authority, authenticity and intentionality of the information obtained. When faced with a long list of search results, young people have trouble evaluating the relevance of the materials presented and often print pages after a superficial glance.

- Mismatch between prior knowledge and the diversity of sources.

- Little reflection on the means of communicating results in accordance with the intention or the context, and a lack of awareness of the ethical aspects involved in information access and use.

- Search engines are the starting point for almost all information searches, not library portals, and most students are happy with their general experiences of using them, because they are better suited to their lifestyles than physical or online libraries.

- Books are the main image associated with libraries, despite the massive investment libraries have made in digital resources. Indeed, most students are unaware of the digital resources that libraries have.

When questioned, librarians stated that undergraduate students do not know how to search in library catalogues or holdings; they do not master the potential of advanced search systems; they do not know how to interpret the reference of an article or journal, perform database searches or evaluate the quality of websites. They stick to their lecturers' electronic dossiers; these play a determining role as bridges or links. Reworking of information is poor, writing processes for different contexts and types of work are not mastered, there is too much copying of information, no thought is given to the organisation of information and there is a lack of awareness of the ethical issues connected with copying and citation.

All of these comments lead to the conclusion that being a digital native is no guarantee of competence, and that work needs to be done with them to attain it. It may be the case that more and more students are arriving at university with fewer information skills due to the impulsiveness, fragmentation and superficiality of information consumption and use. It is very important to raise awareness of the importance of this competency. Librarians also need to be made aware of the need to get closer to these users in ways that allow them to connect.

3.2. Lecturers

Even though it can be assumed that lecturers are information competent, given the high degree of specialisation they have in their teaching or research fields, a few clarifications do need to be made. Lecturers also suffer from information overload, either generally or in their specific fields. They also find it hard to refresh their digital skills when faced with new search systems, new sources of information, new information management software, and new means of communicating knowledge

3. But How Do Students and Lecturers Access and Use Information?

and of taking part in social networks. Information competency is evolving all the time; tools sometimes change the form, pace and moment of academic writing; they imply a revision of values connected with the channels of knowledge publication and dissemination.

Several studies on lecturers' use of information have drawn certain conclusions that are of great concern to libraries. Libraries should act to stem these issues and turn them into opportunities. Even though lecturers' behaviour is different from students' for certain knowledge areas, reports like Ithaka (Housewright & Schonfeld, 2008) show that lecturers also like to find information directly through Google Scholar and other online sources rather than library portals. This implies that a library is somewhat invisible, despite the fact that it is usually the provider of access to many resources that are found through those channels. Lecturers believe that they depend less often on libraries for their teaching and research as the use of digital resources increases; they value a library's role as a buying agent over and above other functions that librarians prefer, such as being the point of access to information resources.

These are just a few examples, but we believe that lecturers may be using information poorly and possibly need to refresh their competencies in the light of new products or the potential of the world of information. It would be rather inconsistent for them not to embrace the lifelong learning and literacy models that they preach to others. Throughout their academic lives, lecturers combine teaching and research activities with management activities that may prevent them from being up to date at all times. That is the reason why they need to refresh their information skills. They need to do this for themselves so that they can encourage their students to do the same.

4. WHAT PREVENTS LIBRARIES FROM HAVING A GREATER IMPACT ON INFORMATION COMPETENCY ACQUISITION?

Given that information is so vast and so complex, that it is accessed and distributed through so many channels, and that it is hard to master and to keep up with; and given that lifelong learning is a requirement that implies being able to draw on meaningful information throughout one's life, there are two questions that need to be asked. First, what obstacles have existed – or still exist – that prevent libraries from cooperating more with students? And second, what can be done to get lecturers to consider this a priority, so that a joint effort can be made to raise the profile of libraries?

The first, as we have already mentioned, refers to the risk of libraries being somewhat invisible, which may affect the expectations that people have of them. In this respect, libraries are called upon to take promotional and marketing actions: they must get closer to users, have a greater presence in their learning spaces, adapt to the different teaching and research needs and habits of lecturers in each discipline, get involved in innovative educational experiences, take part in social networks and means of informal learning of the type that new users like. Specifically, there is still a certain lack of awareness of the information literacy concept among lecturers, students and some librarians even. In comparison with the simplicity of search engines, users feel that library tools and technologies are rigid and hard to use, and this discourages them from using them. We do of course believe that any divergence between users' and librarians' technologies, desires and practices should be avoided. A mutual coming together needs to be achieved in order to facilitate new working processes, channels and better use of information.

Another difficulty has been the slow pace of change in teaching culture, which has held back the implementation of teaching methods that foster a broad, reflexive, critical and intentional use of libraries' scientific

4. What Prevents Libraries from Having a Greater Impact on Information Competency Acquisition?

information, collections and digital resources. Ten years on from the Bologna Declaration, teaching culture is beginning to change, but there have been a number of counterproductive elements hindering that change, such as very little recognition of teaching with respect to research, the lack of activity planning for teaching-learning through problem-solving and library use, the minimal value placed on educational training in the teaching sector and the reproduction of practiced or received methods. In general, the various disciplines have been viewed as a closed set of pieces of knowledge that needed to be conveyed or transferred to students in order to incorporate them into the paradigm in force via a synthetic representation contained in a manual or basic selected texts. In fact, libraries continue to fill up around exam time with students who are prepared to memorise content in shifts, 24 hours a days, 7 days a week. Their study materials are on a virtual campus instead of in photocopiers, the basic texts for the exam are in electronic dossiers instead of in a library, a great deal of information is on the Internet and lecturers project their presentations or web pages in class to give examples of what they are trying to get across in their lectures. To a large extent, however, students still listen so that they can regurgitate pieces of knowledge in a conventional exam. This is why an insistence on supporting the change in teaching culture is still necessary.

With regard to carrying out training activities in libraries, which is very common in all universities (REBIUN, 2008b, c), there are a number of aspects that need to be improved to ensure that they become more successful (Somoza & Abadal, 2007; Roca, González & Mendoza, 2006):

- Educational training for librarians, because training tasks have usually been carried out by specialists in specific scientific areas. Librarians have not been trained to teach or to design instructional activities. It has not been part of the degree in library science and documentation and, therefore, has not been systematised in the university degree curriculum for librarians, though it has gradually been incorporated into professional refresher training plans. As it becomes more widespread, librarians will gradually acquire teaching competencies and the degree of reticence about a duty that they have not traditionally had will be overcome.

- Instructional designs have not been based on students' levels of prior knowledge. Instead, training content has been defined in a more intuitive way, based on the course students were on or a superficial assessment of needs discussed with their lecturers. The lack of awareness of user profiles and their specific needs in the thematic area of their learning prevents more tailored and flexible training from being offered. A minimal, initial diagnostic evaluation of the target audience for the activities has not usually been carried out. An evaluation of the results has not usually been done either.

- The integration of these activities into the curriculum has not been good enough for them to have an impact, to be recognised or to be properly situated. If training activities are not done when and where (the curricular context) the need to search for information arises, or if they are not linked to subjects, then effort and motivation with respect to such activities is diminished. This is a common and serious problem, because the relevance of doing so is not appreciated.

- In training content, instrumental skills (using the library catalogue, databases and sources of specialised information) have prevailed over more conceptual content, such as selecting information available on the Internet, citation methods and the organisation of information. The predominant methodology used has been expository, which is not consistent with the rationale behind information literacy.

- Attendance for advanced training activities (accredited or extracurricular) is usually very low in comparison to the huge success of training activities carried out to welcome new students. This would indicate that they are not linked well enough – from the students' point of view at least – to their academic interests, or that students do not see any benefit in them. It is the students who have

5. Advances, Opportunities and Strengths for Information Competency Teaching in Libraries

attended training activities that are precisely the ones who are more aware of the need to acquire information competencies, not those students who have shortcomings.

With regard to the organisational aspects of information literacy services:

There has been a shortage of human and financial resources to incorporate these competencies, particularly in small libraries. When overloaded with duties, information literacy may be put on the back burner; if it is a priority issue, then the organisation needs to adapt to it. Libraries with a greater awareness of this issue have systematic information literacy plans and reflect this in their organisation charts.

The new role of librarians has not received sufficient recognition. Today, librarians are advisors and consultants on the utilisation, use and relevance of resources that the community can use, which makes their involvement in teaching very variable. Sometimes they take part actively in doctoral programmes when academic rules do not contemplate such involvement, and they teach practically all of the programmes of some subjects outside their working hours, with or without any financial recognition. However, this is something that largely depends on the environment's motivation and predisposition.

Even though libraries are now more recognised by teams of deans and rectors, institutes of education science or IT services, more institutional support is still needed, either that or an overall policy to develop information literacy in a generalised way for all degrees.

5. ADVANCES, OPPORTUNITIES AND STRENGTHS FOR INFORMATION COMPETENCY TEACHING IN LIBRARIES

The above-mentioned difficulties do not undermine the fact that significant advances have indeed been made in recent years. In reality, libraries have always carried out user training in a more or less explicit way: through user reference and consultation interviews, introductory sessions for new students, on-demand bibliographical instruction on specific resources and sources of information, the publication of explanatory guides and so on. From the late 1990s, (Gómez Hernández, 2000), these were gradually extended to deal not only with resource-use skills, but also with more complex competency-related issues. Today, all libraries offer information literacy programmes containing basic or advanced activities – whether self-directed or included in teaching programmes – with a wider variety of content. Libraries also offer activities for teaching and administrative staff, courses on virtual platforms, tutorials available on library websites, face-to-face or online (via e-mail or chats) consultation-response services, collaboration with lecturers on theory lectures or practicals, collaboration with students on their final projects, virtual campus courses, Social Web activities1 and so on. These offerings are highly valued by those making use of them, and librarians are becoming more and more involved with fewer reservations.

In recent years, libraries have also taken a number of organisational decisions to consolidate this service, such as: considering the inclusion and recognition of the librariantrainer in library organisation charts; incorporating this service into the priority lines of strategic plans, institutional evaluation processes and service charters (Roca, González & Mendoza, 2006); modifying and adapting spaces in libraries to create training rooms equipped with computers, projectors and other teaching resources; strengthening relations with governing teams and deans' offices of universities in order to include library presentations or content in different teaching spaces and times; and training working groups.

The EHEA has created a favourable climate for information competency because it recognises it as a generic issue in white papers and in new degree curricula, fostering learning methods that effectively help it to be transferred to all disciplines. Vice-rectors' offices for EHEA affairs and institutes of education science grant institutional subsidies to improve teaching; there are technological resources; educational training is supported; and, after the initial questioning of the Bologna Process, we believe that teaching culture is evolving. A number of ways to include regulated information competency are being considered, and lecturers are requesting the collaboration of librarians. Libraries have an ever stronger presence on virtual campuses, social networks, new student welcome days and university career fairs. In subjects that libraries take part in, services are better oriented and adapted to users' needs and practices, and there is greater cooperation between different services connected with generic competency learning. Thus, relations between language services, career guidance centres, institutes of education science, educational psychology guidance services and IT services are becoming stronger.

6. WHAT CAN BE DONE TO KEEP MOVING FORWARD?

It is logical that institutions like universities should change slowly, and it is essential to carry on working together towards lifelong literacy. For example, it is necessary to maintain and strengthen a knowledge of practices – and to adapt to the practices of digital natives – in order to develop new educational strategies for trainers. This implies that librarians should learn to use the potential of the Social Web's participatory technologies, such as wikis, blogs and social networks, to foster information competencies in informal learning contexts that, today, are becoming an integral part of university students' lives. In a context of information overload, attracting the users' attention in their environments is very important, as is knowing what they want and need, and how to deliver it to them.

It would be a significant step forward if information and digital competencies were a specific and compulsory subject for every degree. This has been achieved in some universities with departments of library science and documentation, where their libraries are well-positioned. This is the case at Carlos III University in Madrid. However, most degree courses are now shorter and departments are interested in holding on to as much teaching as possible. Together, these two factors will probably prevent this model from spreading to other universities. Therefore, as a basic competency that students will have to attain, this competency will need to be situated and related to the specific content of different subjects, final practicals and final projects. It is anticipated that information literacy will be attained by articulating the endeavours of lecturers in their particular subjects and helping them to include information literacy content. Librarians will provide support through complementary courses, tutorials, teaching materials, e-learning or blended learning courses, and work guidance in libraries and virtual spaces (Area, 2007). This will be a great opportunity for librarians to become learning mediators in cooperative environments, a role that, in Spanish, we have termed "entrenauta" (Gómez Hernández, 2008).

This interaction with lecturers needs to be developed at all levels. Librarians should consider them as allies and sources of mutual support by:

- Helping them to keep up to date by providing individual assistance, depending on demand, relating to the sources and tools that they need, at the time and place of their choice. This will ensure that students learn about and subsequently work with these sources.

- Using informal approach mechanisms that work, even though their physical use of the library may be lower due to electronic access: regular talks, such as "technology cafés".

- Providing them with teaching materials and ideas that make it easier for them to work on information literacy in their particular subjects; suggesting joint practicals, offering evaluation criteria or offering to carry out evaluation directly.

6. What Can Be Done to Keep Moving Forward?

- Offering librarians' collaboration in the organisation, training and evaluation of final projects, from the viewpoint of bibliographical correction, reviewing sources relevant to the sector, good structuring of the project and so on.

- Offering institutes of education science and other educational services for lecturer refresher training the chance to include information literacy courses in their programmes, particularly on the use of documentary resources for teaching and research.

- Directly offering the courses that libraries run each year in faculties, with academic recognition agreements to ensure that students take part.

- Providing all the technical, material and human resources (in the same as the CRAI or "resource factories") for libraries to produce teaching and learning materials that lead to changes in traditional teaching.

- Helping to define, design and programme the basic information management competency, so that it can be developed and incorporated into the curriculum, with examples like the UPC guide (2008).

- It is also necessary to work together on the question of how to evaluate information competencies and the results of training actions, not so much (or only) for qualification purposes, but rather for fostering metacognitive processes, their application to new contexts and their transfer.

We believe that the good practice recommendations for developing information literacy services adopted by REBIUN (Spanish Network of University Libraries) (2008a) all point in this direction, as do the conclusions drawn from conferences attended by information literacy managers from various libraries (REBIUN, 2009). In addition to these, we feel that it is important to:

- Try and obtain external accreditations for information literacy training programmes, which are accepted and valued by future employers.

- Cooperate with secondary schools to ensure that pupils arrive at university with an information competency base.

- Train librarians in new teaching methodologies, in an attempt to motivate them to face up to the challenge of competency training and to become part of interdisciplinary teams alongside IT specialists and lecturers in a confrontation-free manner.

- Integrate digital competencies and information competencies, something that we feel is logical so long as the instrumental components do not displace the reflexive and critical components of training (REBIUN, CRUE-TIC, 2009).

- Use 2.0 tools, websites and social networks, and be prepared for Web 3.0, all the while bearing in mind that they are a means to an end and not an end in themselves, on the basis of a plan that gives them meaning.

- Not to leave aside face-to-face sessions, since the concept of information literacy does not imply virtuality, and attempts should be made to ensure that they are very practical and active.

- Try to be useful allies of teaching staff; if they come to a library to seek a solution to a problem, it will be much easier to collaborate on information literacy.

- Carry on promoting the new image of libraries and communicating their initiatives in this sector.

To give an example of libraries that have systematised information literacy teaching, we would mention the library at the University of Seville (2009), because it has a description of its training offerings in basic, intermediate and advanced subjects for undergraduate and graduate students. In addition, it synthesises and integrates models developed by universities like the Open University of Catalonia (UOC), Pompeu Fabra University, Rovira i Virgili University, the University of La Laguna, Pablo de Olavide University, Carlos III University and the UPC. The UPC programme is very strong (UPC, 2007). It has training subjects and activities for undergraduates, final projects, graduates, teaching and administrative staff, subjects for face-to-face learning and virtual campuses, evaluation criteria for these subjects, teaching methods and teaching activity proposals. In the University of Seville's case, we would underscore the introduction of the philosophy of Social Web participation in the development of this training, with blogs and wikis for lecturer training and support, and the production of guides that are cooperatively updated and completed. Another case worthy of note is the library at the University of Granada, which also manages trainer training courses via its virtual campus. The concern for educational consistency is demonstrated by the teaching methods being considered for these activities: the use of a portfolio as a student learning and self-evaluation method, tests by module, discussion forums, coursework related to other subjects, research diaries, practical exercises and so on.

6. FINAL REFLECTIONS

Libraries are making a considerable effort to develop information literacy services, through self-directed learning (while trying to situate the objectives, tasks and levels with students) and cooperation with lecturers, so that the latter can incorporate content into their teachinglearning activities that contributes to students' information competency. It is a long and arduous process because of the slow pace of change in university culture and a considerable number of determining factors ranging from the levels that students have on arrival at university to the characteristics of scientific information in the different disciplines and the confluence of interests in university organisations. Today, several years after the implementation of new degrees, the recognition of a basic competency to use information efficiently – a competency that is connected with others – may lead to a greater integration of libraries and their information literacy services into teaching processes. Faced with an apparent "disintermediation" with regard to accessing and using information flows, information literacy has become one of the main services that libraries are able to offer. In order to attain the objectives of information literacy when learning activities are undertaken, training should be planned by taking account of the students' prior knowledge, the students' practices and activities and the students' needs. Cooperation with lecturers and an evaluation of results are also necessary. It would be unacceptable for university students to complete their education without attaining information competency, since it is a prerequisite for lifelong, cooperative and selfdirected learning. Managing to deliver a higher education that measures up to the demands of the 21st century is a challenge for librarians, lecturers and students alike.

BIBLIOGRAPHY

1. AREA, Manuel (2007). *Adquisición de competencias en información. Una materia necesaria en la formación universitaria. Documento marco REBIUN para la CRUE* [online document]. <http://www.rebiun.org/doc/adquisicion%20de%20 competencias.doc>

2. BRITISH LIBRARY & JISC (2008). "Informe CIBER. Comportamiento informacional del investigador del futuro" [online article]. *Anales de documentación.* <http://r e vistas.um.es/analesdoc/ar tic le/ viewFile/24921/24221>

3. GÓMEZ HERNÁNDEZ, José Antonio (1995). *La función de la biblioteca en la educación superior. Estudio aplicado de la Biblioteca*

Universitaria de Murcia [online document]. Murcia: Universidad. http://www.tdr.cesca.es/TDR-1107106-134142/ index_cs.html

4. GÓMEZ HERNÁNDEZ, José Antonio (2000). "La alfabetización informacional y la biblioteca universitaria. Organización de programas para enseñar el acceso y uso de la información" [online article]. In: Gómez Hernández, J. A (coord.). *Estrategias y modelos para enseñar a usar la información: guía para docentes, bibliotecarios y archiveros*. Murcia: KR, pages 169-255. <http://eprints.rclis.org/archive/00004672/05/ EMPEUIcap4.pdf>

5. GÓMEZ HERNÁNDEZ, José Antonio (2008). "Las metáforas del mundo de la información, y los bibliotecarios" [online article]. *El profesional de la información*. Vol. 17 No 3, pages 340-342. <http://eprints.rclis.org/archive/00013576/>

6. HOUSEWRIGHT, R.; SCHONFELD, R. (2008). *Ithaka's 2006 studies of key stakeholders in the digital transformation in higher education* [online document]. <http://www.ithaka.org/research/Ithakas%20 2006%20Studies%20of%20Key%20Stakeholders%20 in%20the% 20 Digital%20Transformation%20in%20 Higher%20Education.pdf>

7. OCLC (2006). *College Students' Perceptions of the Libraries and Information Resources* [online document]. http://www .oc lc.org/r epor ts/pdf s/ studentperceptions.pdf

8. PASADAS UREÑA, Cristóbal (introduction, translation and adaptation) (2006)."Hacia una universidad alfabetizada en información según Sheila Webber y Hill Johnston" [online article]. *Boletín de la Asociación Andaluza de Bibliotecarios*. 2006. No 84-85, pages 47-52. <http://www.aab.es/pdfs/baab84-85/84-85a4.pdf>

9. PEW INTERNET & AMERICAN LIFE PROJECT (2009). *Generations Online in 2009*. <http://www. pewinternet. org/~ /media//Files/ Reports/2009/PIP_Generations_2009.pdf>

10. PINTO, María; SALES, Dora; OSORIO, Pilar (2009). "El personal de la biblioteca universitaria y la alfabetización informacional: de la autopercepción a las realidades y retos formativos". *Revista Española de Documentación Cientíica.* Vol. 32, No 1, pages 60-80.

11. PINTO, María; CORDÓN, José Antonio; GÓMEZ, Raquel (2010). "hirty years of information literacy (1977- 2007). A terminological, conceptual and statistical analysis" [online article]. *Journal of Librarianship and Information Science. Vol.* 42, No 1, pages 3-19. <http://lis.sagepub.com/cgi/content/abstract/42/1/3>

12. REBIUN (2007). II *Plan estratégico 2007-2010* [online document]. <http://www.rebiun.org/doc/plan.pdf>

13. REBIUN (2008a). *Guía de buenas prácticas para el desarrollo de las competencias informacionales en las universidades españolas* [online document]. <http://rebiun.org/export/docReb/guia_buenas_practicas.doc>

14. REBIUN (2008b). *Materiales didácticos. Competencias en información y habilidades instrumentales* [online document]. 2007. <http://www.rebiun.org/doc/competencias_ habilidades_ generales.xls>

15. REBIUN (2008c). *Programas de formación de habilidades. Bibliotecas universitarias españolas* [online document]. 2007.

16. REBIUN (2009) *Jornada de trabajo de responsables de ALFIN en las bibliotecas universitarias españolas.* <http://www.rebiun. org/export/docReb/jornada_ trabajo_alin.doc>

17. REBIUN & CRUE-TIC (2009) *Competencias digitales e informacionales en los estudios de grado.* <http://rebiun.org/export/docReb/documento_ competencias_informaticas.pdf>

18. RESEARCH INFORMATION NETWORK (RIN), CONSORTIUM OF RESEARCH LIBRARIES (CURL) (2007). *Researchers' use of*

academic libraries and their services [online document]. <http://www.rin.ac.uk/iles/libraries-report-2007.pdf>

19. ROCA, Marta; GONZÁLEZ, A.; MENDOZA, M. (2006). "La formació d'usuaris i les habilitats informacionals: elaboració del pla estratègic de les biblioteques de la UPC" [online article]. APRÈN 2010, page 9. <http://bibliotecnica.upc.edu/apren2010/alin.pdf>

20. SOMOZA-FERNÁNDEZ, Marta; ABADAL, Ernest (2007). "La formación de usuarios en las bibliotecas universitarias españolas" [online article]. *El profesional de la información.* Vol. 16, No 4, pages 287-293. <http://eprints.rclis.org/archive/00011294/01/ epi2007164.pdf>

21. UNIVERSIDAD DE SEVILLA. BIBLIOTECA (2009). *Las competencias informacionales (CI) en las titulaciones de grado y postgrado de la Universidad de Sevilla. Propuesta de integración* [online document]. <http://formacionbus. pbworks.com/ f/ Propuesta%2BIn tegracion%2BALFIN%2Ben%2BTitulaciones-1.doc>

22. UNIVERSITAT POLITÈCNICA DE CATALUNYA. SERVEI DE BIBLIOTEQUES I DOCUMENTACIÓ (2007). *Proposta d'integració i formació de la competència transversal en Habilitats Informacionals (HI) a les titulacions dels estudis de grau i postgrau de la UPC. Document de treball intern.* Barcelona: UPC.

23. UNIVERSITAT POLITÈCNICA DE CATALUNYA. INSTITUT DE CIÈNCIES DE L'EDUCACIÓ (2008). *Guia per desenvolupar les competències genèriques en el disseny de titulacions: ús solvent dels recursos de informació.* Barcelona: UPC.http://www. rebiun. org/doc/formacion%20en%20 habilidades.doc

CHAPTER 10

Academic Education in Library and Information Management in Bulgaria

Rositsa Krasteva[1], Tereza Trencheva[2], Sabina Eftimova[2], Tania Todorova[2]

[1]*Computer Science Department, State University of Library Studies and Information Technologies, Sofia, Bulgaria*
[2]*Library Management Department, State University of Library Studies and Information Technologies, Sofia, Bulgaria*

ABSTRACT

The purpose of this publication is to present the contemporary aspects of training educational and qualification degree "Bachelor" of Specialty "Library and Information Management" of the Library Management Department at the State University of Library Studies and Information Technologies (SULSIT) in Sofia, Bulgaria.In view of specificity and completeness of the presented information in this publication there is a limit, which refers to the training only in educational and qualification degree "Bachelor". The following methods are used: a study of the curriculums of many universities worldwide teaching in this or a related specialty; comparative analysis; synthesis of the obtained information.

Accent is put on the disciplines Intellectual Property, Standardization in Library Activities, Quality Management in Library and Information Activities, Library Psychology and Bibliotherapy. The research draws attention to some aspects insufficiently covered in the curriculum of the programs preparing future highly knowledgeable, trained library and information managers, and offers some solutions, based on our experience in the State University of Library Studies and Information Technologies, to the attention of the LIS academic and professional community.

KEYWORDS

Library Management, Information Management, Higher Education, Bachelor's Degree, SULSIT, Bulgaria

1. INTRODUCTION

Socio-economic and political transformations in the Bulgarian reality necessitated a change in the functions of libraries and enhanced their role in the process of accession to the European and World educational, cultural and economic space. Bulgaria is member of the EU since January 1st, 2007. The libraries in accordance with the principle of wide and equal public access to library and information resources, and as a democratic social institutions contribute to social stability, preservation and development of the spiritual and scientific potential of society. The contemporary challenges facing the libraries, suggest the application of adequate library management. It evolves against the background of changing the overall management paradigm—a process characteristic of global management. The outdated management approaches to the concept of "Rational Management" shall be replaced by the Marketing

1. Introduction

Paradigm, characterized by adaptability and flexibility of management. This undoubtedly imposes new requirements, including management knowledge and skills to the competencies of young specialists, finishing their high degree in specialties like Library and Information Science.

Evaluating LIS Education in Europe, the experts of the European LIS Curriculum Project conclude that in Europe, almost all library and information (LIS) programs in higher education have been developed and are offered within the context of a nation-state [1] . Several authors reported the need for internationalization of library education in Europe in accordance to achieve greater results in student and teacher mobility [2] -[4] . The situation is summarized in LIS Education in Europe: "The Structure and contents of LIS courses vary very much between the different types of LIS education providers in Europe, which include many fairly small academic environments. The apparent disparate nature of LIS educational programmes in Europe constitutes a barrier to increased cooperation in the field. There is a marked need for joint discussions of the structure and contents of LIS school curricula and for identifying and discussing possible common curricular elements both for the purpose of enhancing the quality of individual LIS educational programmes and for the sake of increased collaboration between European LIS school programmes" [5] . Based on these findings, in 2005 started the development of an international project LIS education in Europe, coordinated by the Royal School of Library and Information Science, Denmark. Main goals of the project are:

- To create better opportunities for European mobility of students and teachers;
- To increase the scale of mobility and inter-institutional cooperation;
- To develop a common conceptual framework to define the core of mandatory elements within the LIS curriculum as a basis for increasing the mobility of students and teachers and the acceleration of the Bologna process;

- To work for greater flexibility, transparency and comparability of curricula.

The results of this project were published in the European Curriculum Reflections on Library and Information Science Education [6] . In a separate section was provided information related to the training of library management. Focus is on disciplines such as Marketing Information Services; Communication Skills and Negotiations; Intellectual Property and Information Law; Quality Management and etc.

Taking into account the achievements of our colleagues in Europe, in this article we would like to share our experiences, best practices and suggestions on training in the undergraduate program "Library and Information Management" at the State University of Library Studies and Information Technologies in Sofia, Bulgaria. In response to the social and professional needs, in 2008 in the structure of the Faculty of Librarianship and Cultural Heritage (FLCH) in SULSIT is created new Department of Library Management. Founder and first head of the department was Prof. DSc Ivanka Yankova (from 2011 she is a Dean of FLCH). Since June 2011 up to now— Assoc. Prof. PhD. Tania Todorova is a Chair Holder of Library Management Department. The main task of the Library Management Department is the preparation of bachelors, masters and doctoral students in the field of library and information management.

2. GOAL, RESEARCH TASKS AND METHODOLOGY OF THE STUDY

The purpose of the study is to present the contemporary aspects of training in educational and qualification degree "Bachelor" of the Specialty "Library and Information Management" of the Library Management Department at the State University of Library Studies and

Information Technologies (SULSIT) in Sofia, Bulgaria and to make comparison with the curricula of some of the biggest universities worldwide, which are accredited in LIS education. In view of specificity and completeness of the presented information in this publication there is a limit, which refers to the training only in educational and qualification degree "Bachelor". To achieve the objective are set out the following major research tasks, outlining the methodology of the study: 1) Presentation of the Library Management Department in SULSIT (Bulgaria) and its activity; 2) Exploration, presentation and comparative analysis of the curriculums of many universities worldwide teaching in this or a related specialty; 3) Conclusions and recommendations. The methodology for achieving the objective of the study and solving the set research tasks include the following specific methods: method of study and content analysis, comparative analysis; synthesis of the obtained information. Accent is put on the disciplines Intellectual Property, Standardization in Library Activities, Quality Management in Library and Information Activities, Library Psychology and Bibliotherapy. The research draws attention to some aspects insufficiently covered in the curriculum of the programs preparing future highly knowledgeable, trained library and information managers, and offers some solutions, based on our experience in the State University of Library Studies and Information Technologies, to the attention of the LIS academic and professional community.

3. INTERNATIONAL COOPERATION OF THE LIBRARY MANAGEMENT DEPARTMENT

The creation and development of the Library Management Department and proposed programs—Bachelor Program "Library and Information Management" and Master Program "Library, Information and Cultural Management" are influenced by the principles and guidelines of the

Bologna Declaration and the study and application of leading experience of European and foreign universities.

In the period 2009-2013, the Library Management Department was a partner in two Erasmus Intensive Programmes with the participation of European universities. The Project Das Grimm-Zentrum-(k)ein Bibliotheksmärchen) (2009-2011) with coordinator Humboldt University in Berlin (Germany) created a unique opportunity for cooperation with University of Vienna (Austria), Vilnius University (Lithuania), Masaryk University Brno (Czech Republic) [7]. In the period 2011-2013 Library Management Department at SULSIT was coordinator of Erasmus Intensive Programme "Library, Information and Cultural Heritage Management—Academic Summer School" [8]. Project Manager is Tania Todorova—Head of Library Management Department. Our partners are: Hacettepe University in Ankara (Turkey), University of Zagreb (Croatia), University Paris Descartes (France) and University of Szeget (Hungary). The mission of this project is through using a rich methodological toolkit to implement a modern educational process aimed at implementation of interdisciplinary knowledge and skills relevant to the new requirements in the career development of students in library, information and cultural sector and the policy response to higher education and the EU initiative on "New Skills for New Jobs". The main topics in IP LibCMASSare: Library, Information and Cultural Management; Information Literacy; Preservation and access to Cultural Heritage; Digital libraries; Intellectual Property; Information brokerage; Information technologies in libraries, archives, museums and other cultural institutions [9] [10].

In the process of creating and updating the curriculum of the Bachelor Program "Library and Information Management" important was the impact of active participation of professors and students from the Library Management Department in the short-term mobility under the Erasmus Intensive Programmes. On Figure 1 is shown the dynamics of the participation of professors. OnFigure 2 is presented the students participation. It is necessary to highlight that in 2013 was implemented

3. International Cooperation of the Library Management Department

successfully 1 student mobility of an Erasmus Summer Semester in HochschulefuerMedien in Stuttgard, Germany.

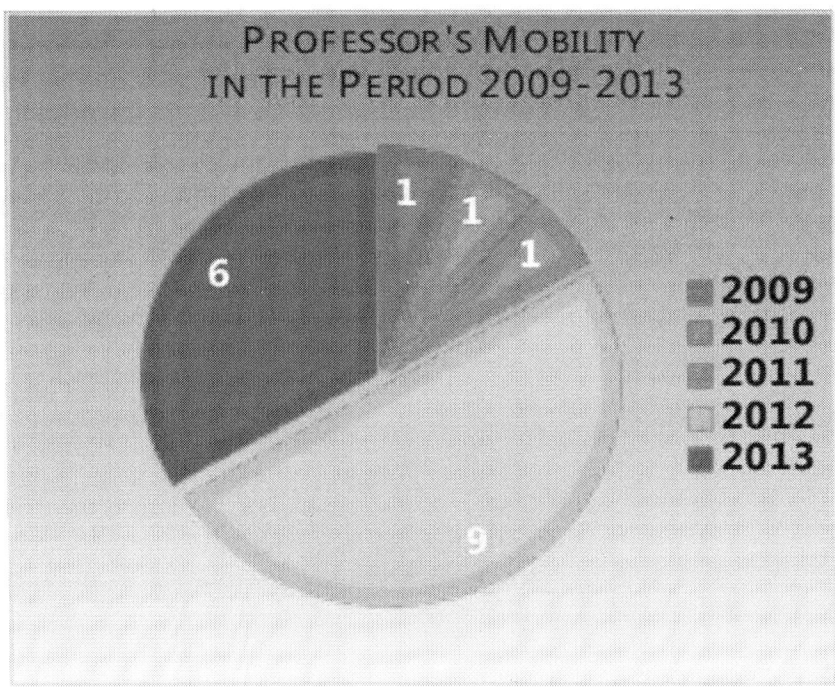

Figure 1. Professor's mobilityfor the period 2009-2013.

These projects were unique opportunity for establishment a stable international network in higher education in Library and Information Sciences, Computer Sciences and Cultural Heritage Sciences and to promote cooperation between academic education and practice—library, information and cultural sector.

On 20th November 2011 the Memorandum for establishment of the UNESCO Chair' ICT in Library Studies, Education and Cultural Heritage' was signed between Irina Bokova, the General Director of UNESCO and Prof. DSc Stoyan Denchev, the Rector of the State University of Library Studies and Information Technologies. Many colleagues from Library Management Department collaborate actively in the management and in various activities and projects of the UNESCO Chair [11].

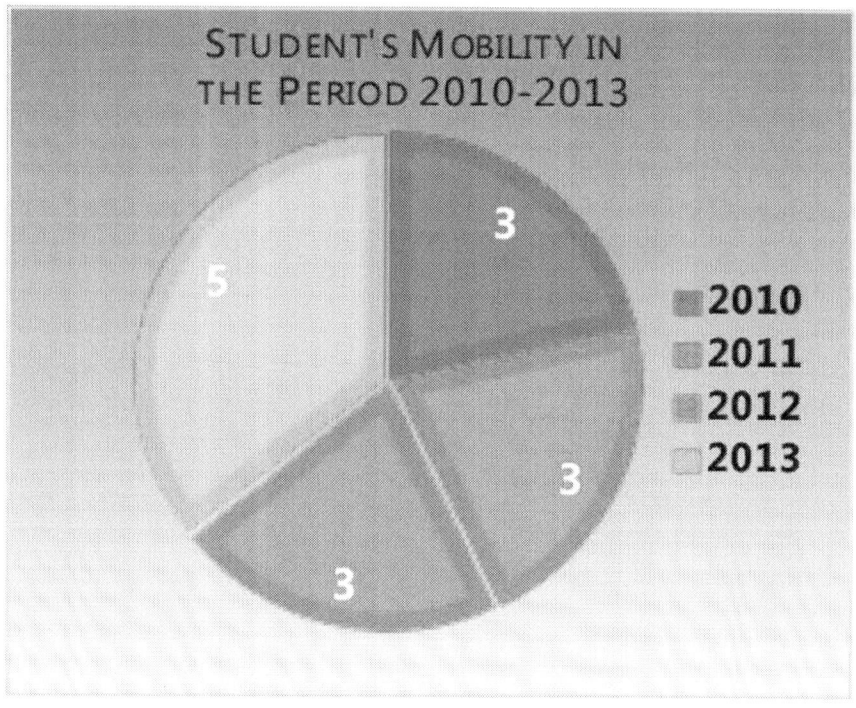

Figure 2. Student's mobility for the period 2010-2013.

We could summarize that for five years period from the establishment of Library Management Department in SULSIT we have achieved a presence in the single educational space in the field of Library and Information Sciences and we are highly motivated for future initiatives and partnerships.

4. CURRICULUM AND METHODOLOGY OF LIBRARY AND INFORMATION MANAGEMENT BACHELOR'S PROGRAM

The students which obtained the educational and qualification degree "Bachelor" in "Library and Information Management" in SULSIT must

4. Curriculum and Methodology of Library and Information Management Bachelor's Program

have a thorough scientific, theoretical and practical training in the specialty, which includes:

- Fundamental training on the nature and the diversity of libraries and other cultural institutions. History of the Field;

- Knowledge and understanding of the library cataloguing, acquisition and bibliographic processes and operations;

- Fundamentals of Library Management, Knowledge Management, Standardization and Quality Management in Library and Information Activities;

- Thorough knowledge of modern library and information technologies and their application to all facets of Library and Information Products and Services. Digitalization and Preservation of Library Funds;

- Theoretical and practical training in Information Resource Management, Information Retrieval, Assessing of Information Needs, Intellectual Property and Information Literacy;

- Specialized training in Library, Information and Project Management to acquire knowledge and skills in planning, organizing and controlling the overall operations of libraries, information centers, community centers and other cultural and public institutions;

- Basic knowledge of Marketing, Public Relations and Library Psychology for optimal organization of information and the use of methods to interact with audiences for library collections, services and events.

Experts should be able:

- To raise and resolve the duties arising in connection with the organization and ensure optimal use of information resources in libraries and other organizations with arrays;

- To develop and manage projects;

- To develop and implement programs for digitization and creating their own electronic resources;

- To organize and manage various initiatives to interact the various audiences;

- Carry out self-sociological, psychological and other research needed to analyze the quality of the services and the extent of their approval by the users, and to be able to improve the characteristics of information and communication forms used in the library, and to bring new services or forms;

- To collaborate effectively with national and international LIS community [12].

Forms of education in educational and qualification degree "Bachelor" are regular and part time and are implemented in eight semesters. Table 1 presents the disciplines currently included in the curriculum for compulsory training of specialty "Library and Information Management" [13].

4. Curriculum and Methodology of Library and Information Management Bachelor's Program

Table 1. Compulsory subjects taught in the educational and qualification degree "Bachelor" in specialty "Library and Information Management" in SULSIT.

Discipline	Semester
History of the Cultural Institutions and Librarianship	First semester
History and Theory of Culture	
Information Resources	
Annotating and Referencing of Documents	
Information Systems	
Book Science	
Rare and Valuable Library Collections	Second semester
Bulgarian Literature	
English	
Second Foreign Language	
Sport	
National Bibliography	Third semester
Management and Library Collections Development	
Library Management	
Automated Libraries—Part 1. Cataloguing, Classification and Subject Indexing of Information Recourses.	
Automated Libraries—Part 2. Automation Library and Information Systems. OPAC, Union Catalogues, National and International Cooperation. Application of ICT in Libraries.	
Intellectual Property	Fourth semester

Information Services in Libraries	Fifth semester
Innovations for Preservation of Library Funds	
Library Marketing	
Information Literacy—Programs and Models	
Architecture of the Libraries	
Library Psychology	
Practicum in various types of libraries	
Information Management	
Knowledge Management	
Bibliometry	Sixth semester
Standardization in Library Activities	
Public Relations	
Library Legislation	
Quality Management of Library and Information Activities	Seventh semester
Digitization and Copyright	
Academic Writing	
System Analysis	
Project Management	Eight semester
Practicum and Fieldwork. Development of Bachelor Thesis.	

*Teaching English and sport is foreseen as a compulsory subject for each semester.

4. Curriculum and Methodology of Library and Information Management Bachelor's Program

There is an option for the students to choose a discipline according to their interests (elective courses). Among the most popular are: Applied Software, Web Design, Content Management Systems, Intellectual Property in Internet, Libraries and Local Authorities and Communities, European Communication Policy. Common culture among students is supplemented by the chosen of them optional subjects such as: Bibliotherapy, Access to information for people with special needs etc. Methodology of teaching the students is done except to traditional methods such as lectures and seminars, and through interactive and situational methods, application of the approach "learning by doing", "edutainment" and etc., which activates the students' participation. Individual forms of work such as term papers, presentations, individual assignments, communication in e-learning platforms—support the learning process and the current verification of students' knowledge. Library Management Department features with the active involvement of students in research projects and joint research papers for participation at scientific conferences and symposiums.

There are two initiatives of the young scientists in the Library Management Department, which are directed to the students. One of them is the work of students in relation to the Week Dedicated to The International Day of the Book and Copyright Day (23 April), with coordinator Dr. Lubomira Parizhkova, where students from all specialties take part into this initiative and have the opportunity to publish their first scientific research in the thematic research collection [14] . The second initiative is led by Dr. Tereza Trencheva and is dedicated to The International Intellectual Property Day (26 April), where the students have the opportunity to meet specialists from practice in the field of Intellectual Property [15] .

For self experience activities the students of Library and Information Management visited: libraries, museums, galleries, workshops for making paper and books, book fairs (with conducting inquiries), book premieres, meetings with authors, public lectures etc. Also, they participate in the initiatives of promotion of reading—marathon of reading, book crossing movement and in different forms of stimulation of children's reading.

Important project, which is creating the opportunity to use the acquired theoretical knowledge in a real working environment is the National Project' Students' Practices' of the Bulgarian Ministry of Education and Science, implemented under the Operational Programme "Human Resources Development" of the EU. For the period July 2013-July 2014, more than 40 students of Library and Information Management Specialty conducted 240 hours internship in libraries, archives and information centers. Our students are actively involved in initiatives of the Bulgarian Library and Information Association and the Association of University Libraries, also. Our goal is nurture an active attitude towards the library profession and knowledge of the real challenges, problems and achievements in their future professional environment.

5. LIBRARY AND INFORMATION MANAGEMENT SPECIALTY WORLDWIDE

For the purposes of this research we have studied the higher education institutions in the world, which provide training in "Library and Information Management" (LIM). In this study, we aim to gain information in which educational and qualification degree is offered training in the specialty LIM and what are the main components in the content of the curriculum. As a source of information have been used official websites of the respective universities. The results are presented in a synthesized form in Table2

Table 2 shows that the universities conduct training in specialty 'Library and Information Management' mainly in master educational and qualification degree. The exceptions are Hochschule der Medien in Stuttgart, Germany; Peking University and The University of Hong

5. Library and Information Management Specialty Worldwide

Kong, China; Moscow State University of Culture and Arts (Department Information and Library Management), Russia; University Technology Mara (Faculty of Information Management), Malaysia.

Practical realization of bachelors graduated in 'Library and Information Management' in Bulgaria, is particularly suitable for libraries in small towns, whose staff is limited (and often reduced to a single librarian) and it is expected to be informed and competent in a wide range of issues. As answer of these needs of library and information sphere, SULSIT offers training in Bachelor level—specialty "Library and Information Management" and in Master level—"Library, information and Cultural Management".

Diplomas and certificates that graduates of the corresponding universities receive are different, account the type of the training already acquired specialty and professional experience and length of the training course.

They can be:

- Bachelour diploma (training lasts 3 - 4 years);
- Graduate sertificate (training is done after obtaining a bachelor's degree or professional experience and usually lasts 6 - 8 months);
- Postgraduate diploma (training is done after obtaining a bachelor's degree or professional experience and usually lasts for 12 months);
- Master (training is done after obtaining a bachelor's degree and usually lasts 16 - 24 months);
- Professional master (training is done after obtaining a bachelor's degree and professional experience and usually lasts 12 - 18 months);
- Graduate entry master (training is done after obtaining a bachelor's degree in Library and Information Management or related specialty and usually lasts about 24 months);

- PhD and other Doctoral Research Degrees (training is done after obtaining a master's degree or completion of a bachelor's degree with special honors and usually lasts 2 - 4 years).

Table 2. Worldwide Universities engaged in training in the specialty "Library and Information Management".

Country	Name of the University	Educational and Qualification Degree	Core Curriculum Elements
Australia	University of South Australia	Master (MA)	- Information Management - Accessing and Organizing Resources - Information Technologies
Canada	The University of British Columbia (School of Library, Archival and Information Studies)	Master	- Organization and Knowledge Management - Information Systems - Digital Libraries
China	Peking University	Bachelor (BA) Master	- Information Storage and Retrieval - Computer Networks - Information Economics - Management Information System - Library Management - Information Policy and Law
China	The University of Hong Kong	Bachelor Master	- Information Science - Science in the Field of Management
Denmark	Royal School of Library and Information Science	Master	- Theory of Production of Information and Knowledge - Analysis and Management of Information Resources - Information Policy and Strategy of the Organization - Planning, Management and Evaluation of Information Systems
Germany	Hochschule der Medien in Stuttgart	Bachelor Master	- Management in Human Resources, Organization, Marketing - Information and Media Market - Library Management System Database
India	Gujarat University	Master	- Management of Library and Information Centers - Leadership and Change Management - Information and Communications Technologies and Research
Malaysia	University Technology Mara	Bachelor	- Information and Library Management - Information, Information Technology and Libraries
Russia	Moscow State University of Culture and Arts	Bachelor Master	- Library and Information Management - Library Statistics, Legislation - Economics and Marketing of Library and Information Services - Information Technology Management - PR and Advertising in the Library
United Kingdom	University of Rhode Island (School of Library and Information Studies)	Master	- Management of Library and Information Services - Information Science and Technology - Types of Libraries and Library Processes and Services
United Kingdom	Robert Gordon University	Master	- Information Studies - Managing Information Services - Database Construction and Use - Knowledge Organization
USA	University of California Berkley	Master	- Information and Communications Technologies - Leadership - Computer Sciences - Psychology and Sociology, Economics, Business, Law, - Library/Information Studies

On the development of educational content of specialty "Library and Information Management" in SULSIT important influence had our joint work in the frame of Erasmus Intensive Programmes with our partners—the leading European universities, engaged in similar training programs

5. Library and Information Management Specialty Worldwide

in the field of LIS education. Table 3 presented the basic information about our partner's programs.

Table 3. Partner universities of the Library Management Department in SULSIT, who carry out training in Library Management and/or Information Management.

Country	Higher School	Specialty and Level	Core Curriculum Elements
Austria	University of Vienna	Library and Information Studies MA	– Management (in Library, Information, Documentation) – System for Information and Documentation
Croatia	University of Zagreb	Library Science BSc, MA, PhD	– Bibliography – Computer Science – Library Management – Information Institutions Management Fundamentals – Libraries and Society
Czech Republic	Masaryk University Brno	Information and Library Studies BA, MA, PhD	– Library Processing and Services – Intellectual Property and Information Activity – Information Science – (Information Security, Information Systems, Information Education)
France	Institute of Technology Paris Descartes	DUT Universal Technical Diploma	– DUT Information Communication – Book Trade
Germany	Humboldt University in Berlin	Library and Information Sciences BA, MA	– Digital libraries – Bibliometry, Infometriya, Scientometrics, – Information Policy, Ethics, Law – Information Retrieval and Exchange of Information – Communication and Management of Information and Knowledge
Germany	Hochschule der Medien in Stuttgart	Information Management BA, MA Library and Information Management BA, MA	– Management in Human Resources, Organization, Marketing – Information and Media Market – Library Management System Database
Hungary	University of Szeged	Library and Information Science BA, MA	– Library Organization, Library System – Management of Technological Resources – Document Management – Legal Institutions of Documentation
Lithuania	Vilnius University, Institute of Library and Information studies	Information Management in Libraries BA, MA, PhD	– Information and Library Service – Information Science – Culture Information and Communication – Creative Communication – Business Information Management – Public Relations
Poland	Jagiellonian University in Krakow,	Management and Social Communication BA, BSc, MA, PhD	– Information and Library Sciences – Culture and Media Management – Electronic Information Processing – Management – Social Policy
Turkey	Hacettepe University in Ankara	Information Management BA, MA, PhD	– Information Management – Information Architecture – Records and Database Management – Information Literacy, Standards and Cooperation in Information Management

Review of the main directions in which are trained future library and information managers can generally be limited to the following: Information and Communication Technology and Resources; Information and Library Management (in particular—Management of Information Services, Management Information Systems, Personnel

Management); Knowledge Organization. Academic subjects included in these directions are different for different universities.

6. FINDINGS FOR FUTURE CURRICULUM REVIEW

A detailed analysis of the information from this research leads to interesting conclusions that will be very useful for future curriculum review of the "Library and Information Management". We achieved some findings and solutions that we suggest to the attention of the LIS academic community worldwide as useful and relevant to the context of modern educational paradigm.

Special attention in this article we would like to devote to the fact that some academic subjects which are included in the curriculum of SULSIT and met student's interest do not exist or are very limited in the curricula of universities abroad. The development and integration of these disciplines in our educational plan is provoked by the interaction with prospective employers that define this knowledge as important and necessary for library and information professionals.

We draw attention to the following subjects:

- Intellectual Property;
- Standardization in Library Activities;
- Quality Management in Library and Information Activities;
- Library Psychology;
- Bibliotherapy.

6.1. The Training on Intellectual Property

In an knowledge based economy, an important place take experts who can interpret issues related to intellectual property, such as librarians and information specialists. It is they who are responsible for creating a policy of promoting understanding and resolving legal disputes and conflicts that are unique to this aspect of the Information Society [16]. One way to achieve this is through the educational impact of the curriculum on intellectual property. Intellectual property can be considered as an element of information literacy in university information environment, so to develop successfully students at the university, and in life, they must learn to use efficiently and effectively the wide variety of information and communication technologies for searching, finding, organizing, analyzing and evaluating the information they need. In addition, they need to understand the ethical use of information, including the violation of individual rights to intellectual property as plagiarism, use without permission of the author of works of literature, art, science, and also of patented inventions, industrial designs, indications (trademarks, geographical indications, domain names, companies). Finally, they should be able to systematize all this knowledge together to create an effective final product. This requires them to assemble the entire package of basic skills for research, technological skills, critical thinking and evaluation.

6.2. The Training on Standardization in the Library Activities

Knowledge of standardization and specific standards in all areas of human activity are not only useful but also necessary in the direct work of specialists and leaders in different organizational structures, including libraries. Learning of this discipline implies a successful career of students and saves them a lot of effort, time and resources to achieve and implement the requirements, rules, standards, approaches, methods and etc., established by the most highly qualified professionals at international, European and national level and involved in the development of these standards.

Among the most common library standards that future library managers in SULSIT learn about can be mentioned:

- BGS ISO 11690:2013 Information and documentation—Library performance indicators.

- BGS ISO 2789:2012 Information and documentation—International library statistics.

- BGS ISO 15511:2012 Information and documentation—International standard identifier for libraries and related organizations (ISIL).

- BGS ISO 5127:2009 Information and documentation—Vocabulary.

- BGS ISO 2709:2011 Information and documentation—Format for information exchange.

- BGS ISO 25577:2012 Information and documentation—MarcXchange.

- BGS ISO 3166-1:2006 Codes for the representation of names of countries and their subdivisions—Part 1: Country codes.

- BGS ISO 11799:2008 Information and documentation—Document storage requirements for archive and library materials.

Standardization in Library Activities discipline, taught in the sixth semester is fundamental for studying in the seventh semester Quality Management in Library and Information Activities.

6.3. The Training on Quality Management in Library and Information Activities

The educational content in this course is structured in two parts:

Fundamental part, which includes the following topics: the importance and relevance of quality management; historical development, total quality management, principles of quality management, principles for analysis, security, control, organization, systems for quality management standards, quality management.

Today more libraries develop and implement systems for quality management. The most commonly used worldwide are those based on a series of international standards ISO 9000 [17]. Therefore lead in the course lectures Quality management in library and information work is devoted to this particular system.Are also presented two other systems for quality management application in libraries—Balanced Scorecard (BSPE) and selforganizations under the criteria of the European Foundation for Quality Management (EFQM).

Specialized part, consistent with the professional field in which to be realized students in "Library and Information Management." This part of the curriculum content provides knowledge, documentation and implementation of systems for quality management in libraries and information centres, in view of their professional activity. The seminars helped to give young professionals skills to formulate the mission of the library, and the ensuing quality policy and objectives to achieve quality. Special emphasis on the brainstorming technique of K. Ishikawaas an opportunity to identify the causes of a problem and their elimination [18]. Students are introduced to the documentation requirements of the quality management system, and sample content of the basic documents of the Quality Management System (QMS), which adapts to the specific type of library.

6.4. The Training on the Library Psychology

The problem of improving the quality of library services at the level of librarians' professionalism today is extremely topical. These factors increase their need for psychological and pedagogical knowledge.

The course on Library Psychology aims to equip students with true understanding and knowledge of the nature, functions and principles of Psychology, as part of the activities in the library environment to provoke new ways of working and communicating with users of library services. The course content is divided into two main modules. In the first module "Library Psychology" is considered library psychology as a scientific and academic discipline, traced the historical development of library psychology, the role of books in modern society. In the second module "Psychology of Communication in the Library" focuses on issues of personal and professional qualities of the library specialist; the psychology of communication between Librarian and Reader; the peculiarities of psychological climate in the library team and theformation steps of an effective team in the library; the nature and style of the library manager. Library Psychology has significant potential, not only theoretically, but practically oriented knowledge useful in the work of librarians from different types of libraries.

6.5. The Training on the Bibliotherapy

The course provides theoretical and practical knowledge to the students about the nature and application of Bibliotherapy in libraries. Students learn the historical and scientific development of Bibliotherapy. The Bibliotherapy brings together at least three fields of knowledge—a Science Studying the Book (Book Science, Literature, Library), the Science of the Human Soul, Targeted Bibliotherapy (Medicine, Psychiatry, Psychotherapy, Rehabilitation) and the Science of reading, (Ensuring Efficiency of Bibliotherapy, Psychology of Reading, Reading

Instruction), which makes it extremely useful in the training of students. The lectures aim to examine possibilities which reveal Bibliotherapy by presenting its various forms, methods and functions. The students learn about new and alternative forms of employment applicable in the library environment.

The course includes the following three aspects: first module "Origin and Nature of Bibliotherapy" which focuses on the scientific development of bibliotherapy, definition of bibliotherapy, determining bibliotherapy, the types and purposes, her functions and tasks. In the second module, "Applications of Bibliotherapy" are tracked innovative techniques used by library professionals, bibliotherapy as a form of working with children with special needs. The third module, "Fairy Tales Therapy" is a brand new in Bulgaria and for the first time available as part of an academic course. In it, students get acquainted with the possibilities of therapy tales methodological model of working with stories, kind words and ways of working with them.

Bibliotherapy is dynamic and flexible science for support, encouragement, challenge, identifying and addressing a number of significant problems in modern society. It is a mechanism for painless reach the soul and the possibility of achieving lasting results and as such may be extremely useful in the training of future specialists.

7. DISCUSSION

In the present article, we pay attention to the above mentioned disciplines, as the analysis of the curricula of many universities around the world has shown that these disciplines are still rarely studied. In Table 2 and Table 3, Universities which studied one or more than one of the subjects on which we focus are present. Relatively new disciplines are presented by us, consistent with the requirements of the modern information society, and the role and functions of libraries in it. We believe that along with the subjects proved important for the preparation of future LIS professionals as library management, library legislation, library marketing, etc. would be appropriate curricula to attend and complementary knowledge of library professionals, consistent with the current state of society and libraries. The information presented can be used by other universities in Europe and beyond for comparison and analysis.

8. CONCLUSION

Through active partnerships with foreign universities, innovative teaching methods appropriate to the training of future specialists disciplines, and the continuous enrichment capacity and experience of the faculty, trained students in specialty "Library and Information Management"—we aim to be one of the leading educational institutions and adequate to the modern needs. We hope that shared insights and achievements will provoke an open professional dialogue for the enrichment of library and information academic programs. This collaboration could be going on together with the LIS community dialogue about IFLA Trends and the future of libraries and librarian profession in general [19].

Acknowledgements This publication has been realized under the project "Development of an information environment to motivate and stimulate young researchers in SULSIT" by Contract: BG051PO001-3.3.06-0055. The project is funded by the Operational Programme "Human Resources Development" SchemeGrantsBG051PO001-3.3.06 "Support for the development of PhD students, post graduate students and young scientists" funded by the European Social Fund of the European Union.

REFERENCES

1. Kajberg, L. (2008) The European LIS Curriculum Project: Findings and Further Perspectives. Zeitschrift für Biblio-thekswesen und Bibliographie, 55, 184-189.

2. Juznic, P. and Badovinac, B. (2005) Toward Library and Information Science Education in the European Union: A Comparative Analysis of Library and Information Science Programmes of Study for New Members and Other Applicant Countries to the European Union. New Library World, 106, 173-186.

3. Georgy, U. (2009) Internationalization of LIS Education within the Bologna Process—Mobility and Flexibility. IFLA, Milano.

4. Kawalec, A. (2014) Education, Competencies, Skills in the Field of Information and Library Science in Europe. Library (R)evolution: Promoting Sustainable Information Practices: Abstracts of the 22nd International BOBCATSSS symposium, 29-31 January 2014, Barcelona, 16. http://bobcatsss2014.hb.se/wp-content/uploads/ 2014/ 01/Bobcatsss-abstract-book.pdf

5. LIS Education in Europe. http://www.iva.dk/LIS-EU/project.asp/

6. Kajberg, L. and Lorring, L. (2005) European Curriculum Reflections on Library and Information Science Education. The Royal School of Library and Information Science, Copenhagen, 241 p. http://www.library.utt.ro/LIS_Bologna.pdf.

7. Pannier, G., Wilhelm, H. and Todorova, T. (2011) IPBib: Das Grimm-Zentrum-(k)ein Bibliotheksmarchen. Mobility and Innovation in the European Context. Proceedings of Evaluation Conference on ERASMUS Intensive Programmes, Federal Ministry of Education and Research, Bonn, 2011, 19-21.

8. Project Website (2013) Erasmus Intensive Programme "Library, Information and Cultural Heritage Management— Academic Summer School". http://libcmass.unibit.bg/

9. Todorova, T. (2012) Library, Information and Cultural Heritage Management: Textbook. Za bukvite-O Pismeneh, Sofia, 246.

10. Todorova, T., et al. (2013) The Changing Role of the Manager in the Digital Era: Findings from Erasmus IP LibCMASS 2012 Project in from Collections to Connections: Turning Libraries "Inside-Out". Proceedings of the 21th International BOBCATSSS Conference on Information Science in Ankara, Ankara, 23-25 January 2013, 202-204.http://bobcatsss2013.bobcatsss.net/proceedings.pdf

11. (2013) UNESCO Chair ICT in Library Studies, Education and Cultural Heritageat State University of Library Studies and Information Technologies. http://unesco.unibit.bg/

12. Yankova, I. (2009) Library Management and New Challenges in Digital Era in Libraries and Their Clients: Free or fee Services Supporting Social Communication in Digital Era. Proceedings of the ePublications of the 15th Jubilee International Conference of JU LIS Institute in Kraków, Poland, 76-80.http:// skryba. inib.uj. edu. pl/wydawnictwa/e06/yankova.pdf

13. State University of Library Studies and Information Technologies.http://www.unibit.bg/learning-activity/bachelor/bachelor-plans.

14. Parizhkova, L. (2013) The Book Our More Sensual Present: Textbook. Za bukvite-O Pismeneh, Sofia, 248.

15. Trencheva, T. and Eftimova S. (2013) Intellectual Property at the Universities—Creativity: The Next Generation. Textbook. Zabukvite-O Pismeneh, Sofia, 206.

References

16. Joint, N. (2006) Teaching Intellectual Property Rights as Part of the Information Syllabus. Library Review, 55, 330-336.

17. Balague, N. and Saarti, J. (2011) Managing Your Library and Its Quality: The ISO 9001 Way. Chandos, Cambridge.

18. Ishikawa, K. (1990) Introduction to Quality Control, 3A Corporation, Tokyo.

19. (2013) IFLA Trend Report. http://trends.ifla.org/

CHAPTER 11

Academic Library Services Support For Research Information Seeking

Jia Tina Du & Nina Evans

Library and Information Management Program, School of Computer and Information Science, University of South Australia, Adelaide SA 500

ABSTRACT

This study investigated the use of a university library academic service to assist in research information seeking, and the role and value of the academic services in support of research from the viewpoints of both academic users and librarians. Ten Ph.D. students completed questionnaires followed by face-to-face discussions and four academic librarians participated in semi-structured interviews. Findings include library online databases and interlibrary loans and document delivery as the services most familiar to, and utilised by, academic users. Assistance in searching for resources in and out of the library, subscribing to the databases and maintaining access, as well as providing training on research skills were major roles and values of the university library as perceived by academic users in research support. Academic librarians believed that users' research-related service needs varied for different stages of the research process and disciplines and they reported perceiving their role as that of a "mediator" and an "information broker". Recommendations for further research are also provided.

In addition to being the key player in terms of teaching and learning support for many years (Sidorko & Yand, 2009), university libraries have developed the function of research support. Paying attention to the information seeking patterns of researchers and satisfying their information needs is crucial for a university library to contribute to the provision of research–related services (Haglund & Olsson, 2008). Many of the recent user-centred library studies have explored specific needs and trends among researchers and discussed the implications for service quality of university libraries (Haglund & Olsson, 2008; Heine, Winkworth, & Ray, 2000; Oakleaf, 2010; Payette & Rieger, 1998). However, limited research has been done that directly explores academic users' library service-seeking and using behaviour in pursuit of their research activities. Understanding the use of library services, meeting the users' service-seeking needs, improving the academic library services provision, and promoting research capabilities and the competitiveness of its university's research are important and associated issues not only for the research in the field but also for the library service practice.

Information seeking and provision are two sides of a coin in terms of the issue of library service support. The seeking and provision of library services work through the interactions between users and librarians. Investigations of both the use and provision of library services can improve the understanding of how the library and its users interact. However, very few studies examine library service support from the perspective of librarians.

The study reported in this article aims to first investigate how academic users utilise the academic library services when seeking research information, including their awareness of library services provision. Second, it examines how academic users perceive the role and value of university library academic services in support of research. Furthermore, it explores academic librarians' perceptions of users' research-related service needs and self-evaluation of service provision. This qualitative research combined different data collection techniques such as semistructured interviews, questionnaires, and face-to-face discussions. This usercentred and perception-focused research requires a more personalised approach to finding information about users' behaviour than could be provided by the mass mail surveys used in many of the traditional library user studies.

1. LITERATURE REVIEW

1.1. Information Seeking on Research Tasks

Internet-based searching has become a major channel for academic users to obtain information for research. Du and Evans (2011) investigated the information searching behaviour of eleven Ph.D. students in an online environment when seeking and locating relevant information for their real-life research topics. The authors identified the academic users' search strategies by analysing the search logs. Due to a research task's explorative and uncertain attributes, academic users usually follow a trial and error strategy by interacting with multiple search systems. Results showed that 45% of the study participants utilised library online databases along with Google/Google Scholar. Moreover, Google and/or Google Scholar were considered as the starting point for most academic information seekers (82%). Academic users dedicate themselves to active information searching through multiple search systems and multiple search queries. Successive searching and obtaining updated information are common requirements for research projects. Results also indicated that accessing and searching the librarypurchased databases was the most used research-related service.

Today's researchers rely heavily on immediate access to electronic information. Networked electronic resources via library portals provide researchers with many benefits that could not be derived when physical use of the library was the only option (Martell, 2008). Kibrige and Palo (2000) also found that academic users such as Ph.D. students who work on research projects with narrower subjects prefer to do more comprehensive searches. The authors claimed that such complex searches towards finding relevant information can be frustrating unless the high expectations are managed through constant education and support by librarians. However, other observations revealed that some researchers were confident that they could manage on their own. These users had very little contact with the library, and were unsure about the potential value of the competencies and skills of librarians (Haglund & Olsson, 2008). A gap may exist between users' perceptions and academic library's services provision (Du & Evans, 2011).

1.2. The Role of the Academic Library and Librarian

For the purpose of this paper, the terms university library and academic library are used interchangeably. A review of the literature suggests that the most important service provided by the academic library is to offer training programs to improve users' information literacy and assist them to find and use resources and other information (Sidorko & Yand, 2009). Ren (2000) found that researchers' selfefficacy in electronic information searching increased after the training.

There is an increased demand for assistance with electronic information search and use. For example, to make research and publishing easier, the University of Hong Kong library signed an Endnote site licence with the vendor to make training and delivery available to assist academic users in learning to use the software for bibliography management (Sidorko & Yand, 2009). Another important service provision in an on-line searching context is RSS feeds. Academic users can subscribe to academic

1. LITERATURE REVIEW

publishers' digital libraries which offer an RSS feed for each journal and reporting summaries of each new issue as it becomes available, thereby staying current with emerging knowledge in the field (Kim & Abbas, 2010).

Pinto, Fernandez-Marcial, and Gomez-Camarero (2010) explored the opinions of researchers about the services provided by university libraries. They found that the researchers seem to place little value on the role of the librarian in his or her training as an information user. Users' decisions not to use reference services were found to have been made not because they do not need help but rather because many of them do not regard librarians as valid information sources or resources in solving their specific and specialised needs (Pinto et al., 2010: 77). The authors further claimed that university libraries need to improve their relationship with users and understanding users' needs, hopes, and expectations in order to assist innovation in scholarly activities (Pinto et al., 2010).

Lu and Guo (2009) suggested that academic libraries need to pay more attention to users' behaviour and provide more individualised and professional services.

Shumaker and Talley (2010) referred to the model of "embedded librarianship" where information professionals need to enhance their ability to understand users' research issues and needs. Haglund and Olsson (2008) further discussed the role of a university library in contributing to the competitiveness of its university's research and suggested that librarians need to "leave the library building" to work in the research environment and be engaged by the researchers. Barrett's (2005) study has uncovered different needs at different stages of graduate students' programs which provide implications for librarians about services in terms of particular 'zones of intervention': librarians can provide key services in critical stages of topic selection, focusing, and project initiation beyond simply helping with tracking down specific works.

Previous library studies consider the views of users and call for understanding of researchers' information needs and seeking behaviour in order to provide the important service and add value to scholarly activities. However, very few studies have examined university library services provision from the perspective of librarians. The mode and understanding of service provided or displayed by the academic library and librarians may also influence users' service-seeking behaviour during research activities.

2. RESEARCH DESIGN

2.1. Background: The academic library services in the University of South Australia

The University of South Australia (UniSA) is the largest university in South Australia, with 36,156 students and 2,396 staff based on five campuses; four in Adelaide and one in Whyalla (UniSA, 2011). As the leading provider of international education in the state, the University is also a significant contributor to South Australia's economic growth and prosperity. The UniSA library plays a significant role in supporting the University's teaching, learning, and research mission through the provision of relevant information resources and associated services. The library aims to align with the University's online strategy, to complement the University's teaching and learning framework, to reflect the information and learning needs of a diverse student body, to provide global access to quality information resources for Research Centres and Institutes and to support the scholarship of teaching. To achieve this, the UniSA Library underwent a staff restructure in 2004 resulting in a transition from autonomous liaison librarians to a team approach for

2. RESEARCH DESIGN

service provision. The Academic Library Services (ALS) teams were formed in 2005. The four ALS teams consist of professional librarians dedicated to each of the four academic divisions of the University: the Division of Information, Technology, Engineering and the Environment; the Division of Health Sciences; the Division of Education, Arts and Social Science; and the Division of Business. They provide various training programs to support teaching, learning, and research activities mainly by means of individual consultations and workshops.

Individual Consultations (IC) are provided to assist with finding library resources, using electronic information resources, advising on locating journal impact factors and citation data and electronic solutions for keeping up to date with research activities, and recommending databases for research. A program of workshops (W) known as Strategies for Successful Research (SSR): Finding and Managing Information includes core workshops and division-specific workshops. Core workshops are provided University wide, such as Searching beyond Google and Making your research visible. Division-specific workshops are provided with a discipline focus, such as Database searching for your research: Art, Architecture & Design. These are presented both on-campus and online to address researchers' information needs. Other noteworthy research-related services include interlibrary loans and document delivery, assistance with managing references, as well as keeping a collection of research degree candidate proposals. Staff and students can also suggest new resources for purchase by the Library and the progress of commencing Higher Degree by Research (HDR) students is personally tracked.

2.2. Data collection

A sample of ten Ph.D. students and four academic librarians from UniSA participated in the study during September and October 2010. The volunteer Ph.D. students were recruited by individual e-mails describing the purpose of the research. Ph.D. students who were willing to participate in the study replied to the invitation e-mail and were selected on a "first come, first served" basis. An attempt was also made to select study participants from all four divisions/ disciplines to avoid introducing bias due to different information seeking habits. The study participants were invited individually to a meeting room on the City West campus at their convenience.

A questionnaire was presented to the participants who were required to fill in as much information as possible, followed by a face-to-face discussion. The discussions allowed the participants' further clarification and explanation on their answers. These discussions were recorded. Despite the small sample, the valuable face-to-face discussions allowed capturing comprehensive reflections from the participants, which could not have been derived from mass surveys.

A total of 10 Ph.D. students (seven male and three female) from the four academic divisions/disciplines participated in the study (Table 1). The participants ranged in age between 20s and 60s and were in varied study years.

Semi-structured interviews were conducted with four academic librarians from each of the four ALS teams individually in a meeting room on the City East campus. On average the librarians had 10 years relevant working experiences. Each interview lasted around 30-45 minutes and was recorded. The interview questions focused on users' requests, information services provision, initiation and follow-up, and self-perception.

2.3. Data Analysis

The data collected for analysis included (1) questionnaires, including PhD students' awareness, interactions, and expectations of the University library academic services; and (2) audio recordings of interview and

2. RESEARCH DESIGN

discussion with both librarians and students. The interview and discussion recordings were transcribed by a professional transcription company. The questionnaires and interview transcripts were qualitatively analysed by the researchers using the technique of content analysis. The key point of applying the content analysis method is to classify the full text into categories (Weber, 1990). The content analysis of Ph.D. students' questionnaires and interview responses from academic librarians developed preliminary taxonomies of users' awareness, interactions and expectations of the University library academic services, and librarians' perceptions of researchrelated service needs and self-evaluation of service provision.

Table 1. Summary of demographic data

Gender	Number	%
Male	7	70
Female	3	30
Age		
26-34	4	40
35-49	3	30
50-64	2	20
65-74	1	10
Discipline		
Business	4	40
Education, Arts and Social Science	2	20
Health Sciences	2	20
Information Technology, Engineering and the Environment	2	20
Study year		
1st	2	20
2nd	1	10
3rd	2	20
4th	5	50

To protect the participants' identities, each study participant was randomly assigned a number in the analysis, for example, S1 refers to a participating Ph.D. student, and L1 refers to a participating academic librarian.

2.4. Limitations

This study investigated a small sample size of Australian Ph.D. students in a local university as study participants. Further studies to examine users' library service seeking and using behaviour with larger samples and different user groups such as academic and research staff would be beneficial.

3. RESULTS

3.1. Academic users' utilisation and perception of the university library academic services

Although the awareness and use of the library's online information resources was the highest use made of the Library, the students also used the physical collections and services provided on-site.

The purposes for Ph.D. students to physically visit the University library were mostly to find and return books, videos or other materials. It is noteworthy that 32% of the visits were related to interacting with library staff to obtain assistance or in workshops. Fewer Ph.D. students browsed the internet (5%) or studied or read in the library (5%), probably because they could approach these services from their office or at home. To find and return books or other materials were two major purposes for which Ph.D. students visited the University library. S1, S2, S6, and S9 further expected to approach and read the latest and best books with more copies available in the library.

3. RESULTS

Table 2. Purpose of physical visits to the university library

Purpose of physical visits	Number	%
Find books, videos, etc. on specific topic(s)	9	22
Return books or other materials	8	20
Attend a workshop, class or meeting	7	17
Get assistance from a librarian	6	15
Use the photocopy machine or other equipment	4	10
Browse in the library	3	7
Browse on the Internet	2	5
Study or read	2	5

Table 3. Awareness and utilisation of the university library academic services

University library academic services (Note: IC- Individual Consultations; W- Workshops)	Number Awareness of the services	Number Utilisation of the services	Utilisation /awareness ratio (%)
Library online databases (self-service)	10	9	90
Interlibrary loans & document delivery	10	7	70
W- Database searching for research	9	5	56
W- Strategies for searching databases	9	5	56
W- Where could I publish	9	2	22
IC- Using electronic information resources	8	2	25
IC- Finding library resources	8	1	12.5
Managing references – vendor training	8	1	12.5
W- Finding theses	6	3	50
IC- Advising on locating journal impact factors and citation data	6	1	17
IC- Recommending databases for research	5	3	60
Suggesting new resources for purchase	5	3	60
W- Keeping up to date	5	3	60

IC- Using electronic solutions to keeping up to date	5	1	20
W- Discovering and analysing cited papers	4	2	50
W- Measuring quality – journals and publications	4	2	50
Research degree candidate proposal collection	3	1	33
Tracking commencing HDR students	3	0	0
W- Company and industry information	3	0	0
W- Searching beyond Google	3	0	0
W- Searching for evidence based resources and systematic reviews	2	1	50
W- Australian statistics	1	0	0
W- Exploring the medical subject headings	1	0	0
W- Managing and tracking publications for grant/award applications	1	0	0
W- News online	1	0	0
W- Patents	1	0	0
W- Discovering and exploring *Compendex* and *INSPEC*	0	0	0
W- Discovering and exploring SciFinder scholar	0	0	0
Total	130	53	41

Library online databases (self-service) and interlibrary loans & document delivery were the most familiar and most frequently utilised services by the participants. All the participants were aware of the library online databases (self-service). All but one of the participants (90%) used the library purchased electronic databases to search for resources and information on their research topics.

The second most familiar service was interlibrary loans & document delivery. The utilisation/awareness ratio was 70%. Study participants

3. RESULTS

stated "I've used the interlibrary stuff to get a book from Mawson Lakes campus over to City West campus so I wouldn't have to drive out there and get it" (S4) and "If I find there is a conference paper or something that I really want but I just cannot find it anywhere, I contact the library and they've helped a couple of times" (S7).

Two workshops - Database searching for research and Strategies for searching databases were also known and welcomed by research students. Both utilisation/awareness ratios were 56%. One study participant stated "I usually receive some e-mails from the library regarding the workshops and I have attended the database searching one" (S8).

Even though they were not that familiar to users (only 5 of 10 participants realised), the services of recommending databases for research, suggesting new resources for purchase by the library to support research, and the workshop Keeping up to date had a 60% utilisation/awareness ratio. Similar results were found in the workshops Finding theses, Discovering and analysing cited papers, and Measuring quality journals and publications, in which the utilisation/awareness ratios reached 50%.

3.1.1. Reasons for not using the University Library Academic Services

The relatively low utilisation/awareness ratio for many services shown in Table 3 indicates that some participants have either never used certain academic services provided by the University library in assistance with their research, or they did use a service but dropped it at a later stage. The following reasons were found for academic users not using certain services:

- Have confidence in obtaining information themselves

Four in ten study participants preferred to do their own searches rather than delegating them to the librarians because they were quite confident in finding relevant and sufficient information. Study participants stated "most of the workshops were about information searching, I have confidence in finding books or any academic materials needed for my Ph.D. thesis. I don't think I need to attend those workshops" (S1) and "I haven't ever consulted with library people to help find any papers. I can find them. The internet is very convenient" (S6).

- Not sure about the value of librarians in assistance of research

Forty percent of the participants doubted the effectiveness of research assistance from librarians. One study participant stated "there's been a couple of times I've gone to get help from the library and they can't really help me....I want to know how to do advanced searches on Science Direct and they don't actually know so they spend 10, 15 minutes doing what I've already done and haven't been able to work it out myself" (S8).

Some participants were not sure how librarians [individual consultations] can help because "the research topic is too specialised" (S2, S9). They sought help from supervisors or fellow students who were in the same research field. For example, "I consult with my Ph.D. supervisor. I've found that that works better for me" (S4), "there's an expert [supervisor] here to help on my research topic" (S2) and "If I cannot find one paper, I will try to ask some of my colleagues to help" (S1). According to George et al. (2006), assistance from friends and academic staff members influence the information seeking behaviour of research students.

- Have no idea about a specific service

Two study participants did not know a specific service was available in the library. For instance, one study participant stated "in fact I did not know there is such kind of service [individual consultations]" (S1) and "I didn't know they did that for journal articles [document delivery]" (S4).

3. RESULTS

- Waste of time to attend workshops

Some study participants have attended workshops but they did not feel it was worth the time and effort and did not plan to use this type of service again. "One workshop that I did go to didn't move through fast enough. The last bit is where I really got some useful information but having to waste 45 minutes to get maybe five minutes of actual proper interesting stuff " (S4) and "The first hour might be on quite basic stuff which I already know so, to me, I'm not using my time wisely" (S8).

- Have difficulty in using the library online databases

The high degree of complexity and difficulty is an obstacle for using the library online databases. "I forgot how to use them almost immediately even though they are clearly useful. I want to learn how to use the online resources for research but being over 60 it doesn't come naturally to me…I don't do it very well. A lot of options there to refine the results – that's a hassle" (S3). This study participant had given up noting other easier and effective access methods were available.

- Other alternative types of information sources

Web search engines and professional websites were found to be important resources and convenient to obtain information. For example, almost all the participants used Google and Google Scholar when searching for research information. "I use library online databases whenever Google cannot provide satisfactory results" (S1), "All I use is Google and occasionally Google Scholar" (S3), "I prefer to use Google Scholar because I've used it for several years, it's easy to use" (S6), and "I use Google and Google Scholar now a fair bit to verify the article I already had …I found that to be quite useful - more useful than I thought" (S10).

Sometimes professional websites were considered as a good option for obtaining specialised research information. "The IEEE website is another main source for me to get journal articles, conference papers, and other research papers" (S2) and "I also have a website called DBLP which lists all the conferences on the computer and information science…I use this website for searching info on conference papers and downloading electronic version of proceedings" (S1).

3.1.2. Perceptions of the Role and Value of the Library and Academic Librarians

The role and value of the library and academic librarians in terms of research support, as stated by the academic participants, were classified into the following four categories (in order of importance):

1. Assistance in searching for resources in and out of the library

Providing assistance for information searching, including developing and refining search strategies, was considered to be the major contribution made by the library and librarians. For instance, study participants stated "I go to seek help whenever I have difficulty in finding and locating books and journal articles" (S7), "They should help define how you search in terms of the process you go through with the databases. I don't want them to try and tell me what key words to use. I want them to show me how to use the key words" (S8), and "That's an area where we really call on librarians for working out the best search strategy and the databases that are going to be best for finding the largest quantity and quality of references for that" (S10).

2. Subscribing to the databases and maintaining the access

For example, both S4 and S7 emphasised the importance of keeping the subscription to all of the databases to which they currently have access. The library could provide more full-text resources [in the databases] as "currently only an abstract was available for some papers" (S6).

3. Training centre for research skills

Study participants believed that training by librarians on how to use advanced searches in databases should be very helpful as "the search skills and the structure we need to learn from librarians" (S8). They would like clearer directions on saving searches in various databases and skills for setting up follow-up alerts for updated research papers and "understanding impact factors and rankings and how they relate are really helpful" (S10). One study participant would attend workshops as they fit with the research agenda, "where could I publish is relevant when preparing my first publications during PhD candidature" (S9).

4. Being part of the research team

One study participant insisted that the library and librarians be part of the research team. "I wouldn't have been able to do it without them" and "It's the delight of having a librarian show personal interest in my research endeavors, know my research topic, email me out of the blue with a link, article or reference that relates to my research". He or she greatly valued the library and librarians in support of their PhD research "Within the library what I found was a really useful kind of group of people who understood research, who talked to me as if I could be there" (S5).

3.2. Academic librarians' perceptions of users' research information needs and their role of service support

3.1.1. Types of Inquiries Received from Ph.D. Students

- The types of inquiries depend on the stage of the research process.

Commencing students wanted general information from the very basic perspective such as how to use the library catalogue and where to start. "They come along and say this is my proposal, I haven't really done anything; can you point me to where I should start" (L4). Early stage students wanted assistance with finding literature for their literature review and research methodology. As students approached completion, the librarians would receive different queries. Students who have been around for longer and who have done a literature review often wanted further assistance with bibliographic management and how to develop their research. Students at the very end of the chain needed information about topics such as copyright permissions.

- The types of inquiries also depend on the discipline and area of research.

For example, students from Health Sciences conducted systematic reviews of the literature and some of them actually published systematic reviews. Those students were often asked to give a very detailed search strategy and the databases did not always use the same terms. "Medical students do very complicated searches and they tend to get in a bit of a muddle with combining concepts. I've got a series of little tips that I learnt through experience with very complicated searches and showing students how to break it into sections and actually save sections of their search" (L2).

- Common inquiries.

Common requests were for assistance with setting up alerts for keeping abreast of new information. Finding theses was also a common query and the library ran an online range of workshops about this process. Quality measures and issues were main concerns for the Ph.D. students who asked for things such as how to check the quality of a journal, impact factors, and to check whether they were peer-reviewed.

Another common query was about "the stupid technology not working", particularly when trying to access the electronic version of the journals. According to librarians, many students came back to do higher degree research after maybe ten or twenty years. "They're the ones that need a fair amount of hand-holding at the beginning. They have not used a computer often or for a long time or ever, they have no knowledge of electronic researches and this is totally new to them" (L4).

3.1.2. Academic services initiation and follow-up

Inquiries were sometimes initiated by Ph.D. students and sometimes by library staff. E-mailing to the ALS team shared e-mail address was promoted as the preferred method for inquiries or booking individual appointments. Librarians would do some preparation on users' topics before going into the interview.

Library staff also sent bulk e-mails to students about library induction sessions, the available services and facilities and workshop information as well as individual e-mails to students who missed the library induction session or one of the workshops, inviting them to schedule a personal appointment with a librarian.

During the previous year, the library trialled a tracking service for commencing Ph.D. students. Students were emailed to determine whether they have completed everything on the checklist of services provided by the academic library services team, and to ask what assistance they need. For example, they tracked which students attended the SSR workshops, but "it proved to be too much work for too little return" (L1).

With regard to the follow-up service, the librarians had different experiences. For example, one librarian stated "Some of them seek further assistance after the first interaction. Some of them you never hear from again; you have no idea how they're going" (L1). However, another librarian felt that "I'd like to do that follow-up and have that personal approach with my students. I just think that if you can build a relationship then you can build your services around that. If we're not proactive then we're not promoting the library and the services and we're not giving the students everything that they sign up to" (L4).

3.1.3. Librarians' perception of their role in support of research

The librarians believed that the role was very much that of a "mediator" and an "information broker". For example, one librarian stated "We're the person in the middle. We can try and steer them in the direction of where they'll find useful information. We can talk to them about their search strategies; talk to them about the search terms that they use; get them to [use] lateral thinking, synonyms and remember that different databases come from different originators, so terminology can be different" (L4). Also, "To make sure that they look at the databases that cover both of those disciplines. So I think that's our key role, really" (L3).

One librarian personally attended the research seminar series where Ph.D. students presented their preliminary results. "I often attend those sessions, particularly for the students that are new and just getting their proposal up and then they're starting their research and they're reporting back on that. It's really nice to hear what they're doing and then I can then follow that up. I really enjoy that because otherwise I'm just sitting at my desk and it's so sedentary. I just want to get a little bit more out of what I do and hopefully that then is good for the students too" (L4).

This suggestion emphasises the librarians' role of being proactive in service provision but was not necessarily linked to the concept of the 'embedded librarian' as discussed in literature (Shumaker & Talley, 2010). Our results show that the librarians did not believe they should get involved in the student's research analysis, the content or the research topic. "Most of our Ph.D. students actually come with a reasonable idea of what their topic is because they're often working very closely with other research that's going on within the division" (L2).

However, the librarians sometimes did get requests for assistance in doing research toward finalising the research proposal and the analysis. Librarians believed that these requests were more suitable for the research education advisers in the Learning and Teaching Unit. "They are key people who usually have Ph.D.s and can therefore assist students with more than information searches. They run sessions on finding the gap, how to do literature reviews and methodologies that the librarians often did not get involved in." (L1).

Librarians did not really help students develop their research topic. However, librarians can add value to helping refine students' topics if it was clear that they were going the wrong way. "There was one particular one that you could see that what she was trying to do would take her like 10 years to do her PhD. You could sort of flag some things that probably the supervisor would pick up anyhow that she might need to put some constraints on" (L4).

3.1.4. Value recognition

Librarians expressed their desire to be recognised by academic colleagues. "I think we're lucky in this division because the academics also recognise the value of the library. So they will also promote the library services and obviously ask as well. So we have a very good working relationship with divisional academics as well" (L3). One Ph.D. student got to know the librarians because his or her supervisor suggested "if you ever need a book ordered in because it's not available at that stage then it pays to know them [librarians] and to be able to tell them why you need the book, why it is important for your research, and so on" (S9).

Librarians felt valued if users achieved positive results after making use of their services and they received positive feedback from users. User satisfaction seemed to be a high priority for the librarians. "It's so lovely when the penny drops and the nice little e-mails you get back saying, 'Oh this was fantastic'. In fact, I got one from one of the Ph.D. students when we sent out a follow up e-mail at the end of last year. She said 'I don't know how I would've survived the last six months without you'. So it's just little things like that that just make it all worthwhile" (L2).

4. DISCUSSION AND RECOMMENDATIONS

Our results provide insights into PhD students' library service-seeking and using behaviour and their viewpoint of the University library and librarians in support of research. The results also presented academic librarians' perception of users' research-related information service needs and their own perception in support of research.

4.1. Academic users' viewpoint of library and librarians in support of research

Academic users do access the University library physically for finding and returning books or other materials, in addition to using online resources and services. The results are different from prior work which suggests that the use of the physical collections and services of academic libraries continues to plummet while use of electronic networked resources skyrockets (Martell, 2008).

The most familiar and utilised service was Library online databases (self-service), with 90% utilisation/awareness ratio. This is followed by Inter-library loans & document delivery (70%), and the workshop Strategies for searching databases (56%). Researchers were mainly concerned with accessing, searching (including strategies) and obtaining information and resources.

Some services were very familiar to users but were seldom used, such as the workshop Where could I publish (9 of 10 users knew about the service) and individual consultation-Finding library resources (8 of 10 users knew about the service), with utilisation/awareness ratios of 22% and 12.5% respectively. In contrast, other services, for example, Suggesting new resources for purchase by the library and the workshop Keeping up to date, had a high level utilisation/awareness ratio (60%) although only 5 of the 10 users were aware of the services. It indicates

4. DISCUSSION AND RECOMMENDATIONS

that these services were important and popular and they should be known by a wider range of users via some kind of promotion.

In terms of academic users not using the University library academic services for assistance with their research, confidence in obtaining information themselves and being unsure about the value of librarians were two major reasons. Our results confirm and extend the opinion of Pinto et al. (2010) who suggested users were confused over the level of librarian's knowledge and functions. We found that users expected academic librarians to provide knowledgeable and timely services. For instance, "I would prefer them to actually know and they can just show me instead of having to sit there and watch them do what I've already done which is a bit frustrating" (S8). University library academic services are expected to respond with greater efficiency and efficacy in a professional manner.

Due to differences in techno-literacy and searching experiences, the academic users surveyed demonstrated varying abilities and experience related to finding and using resources for research purposes. For users who have not performed much online information searching before, using library electronic sources can be a daunting task. Students who had returned to university to do higher degree research after ten or twenty years had little knowledge of electronic resources and needed some hand-holding at the start and the pace of training workshops was too fast for them. In contrast, young generation Ph.D. students who grew up with the Internet felt the pace of the training workshops was too slow and regarded attending generic workshops as a waste of time.

In order to address the changing information needs of users, library service providers need to understand what influences information searching behaviour and activities. Personal traits, such as experience with the Web, online databases, and information literacy level should be considered when the library provides training workshops. Surveying users to determine their individual behaviour patterns and experience in advance of training allows for librarians to develop better and more directed support for the researchers' endeavours.

With regard to the perception of the library and librarians, four types of research support roles were identified: 1) searching assistance for resources in and out of the library, 2) subscribing to the databases and maintaining the access, 3) training centre for research skills, and 4) being part of the research team. Users presume the University library correlates closely to information searching and acquisition (access to various professional online databases), and advanced search and research skills training.

All but one of the participating Ph.D. students stated the importance of the library and librarians as assisting their research activities in various ways and that the help added value to their work at different levels. During the face-to-face discussions, the participants said they were receptive to good and timely advice and services from the librarians. This may encourage librarians to do more than provide access to information but they could actively participate in research construction as well (Haglund & Olsson, 2008).

4.2. Academic librarians' viewpoint of the library and librarians in support of research

From the librarians' point of view, academic users' research-related information needs vary depending on the stage of the research process and the field of research. Accordingly, the University library provided discipline-specific workshops to adapt to the needs of users. The library also sent bulk e-mails to commencing HDR students about library induction and information about the services and facilities accessible in the library. The library also provided a series of topic-based workshops with the aim to fit in with different requirements along the research process.

4. DISCUSSION AND RECOMMENDATIONS

Regardless of differences in academic users' research stages or disciplines, they were considered to be seeking high quality information. Other common inquiries from Ph.D. students included setting up alerts for keeping up-to-date information, finding theses, and helping solving technology issues when accessing the electronic resources.

Librarians held different opinions on providing follow-up services. Some were proactive while others were relatively passive. Ph.D. students usually do research over a long time span, so the longevity of their research and how to keep them focused should be considered. Users' needs can only be met if they are clearly identified. It is important to understand that users' expectations are dynamic and depend on personal needs (Lu & Guo, 2009). We may suggest librarians need to be proactive by making contact with individual students and personalising the service.

Librarians regarded it to be their role to ensure that academic users know what resources are available, where to find the useful information they need, how to use appropriate search strategies including choosing search terms and search databases; and how to evaluate the information. The librarians did not believe they should get involved in the student's research analysis and be embedded in a specific research area.

The librarians' self-perceptions of their role are similar to the academic users' perceptions of the role that the library and librarians could play. However, academic users' expectations regarding how to be an effective and efficient searching assistant, which databases need to be subscribed to, what types of training are needed for improving search and research skills necessitate that libraries and librarians work proactively and more closely with academic users, although being embedded in the specific research topic is unnecessary.

5. CONCLUSIONS AND FURTHER RESEARCH

This study investigated researchers' library-service seeking and using behaviour. The role and value of the University library and librarians in support of research were also discussed from the standpoints of both users and librarians. Investigations of the use of an academic library's services can inform library management about how the library and its users interact and can assist in a review of library policies. Academic librarians should spend time thinking about the future and work towards a vision where valuable services are provided to its users.

ACKNOWLEDGMENTS

This study was funded in part by the SIM Lab in the School of Computer and Information Science at the University of South Australia. The authors wish to thank the study participants for their time. We acknowledge Irene Doskatsch, Margaret Heslop, and David Evans for their assistance in recruiting study participants at the University. The authors also thank the anonymous reviewers for their comments and suggestions.

REFERENCES

1. Barrett, A. 2005. The information-seeking habits of graduate student researchers in the humanities. The Journal of Academic Librarianship. 31: 324–331.

2. Du, J. T. & A. Evans. 2011. Academic users' information searching on research topics: Characteristics of research tasks and search strategies. The Journal of Academic Librarianship, DOI: 10.1016/j.acalib.2011.04.003

REFERENCES

3. George, C., A. Bright, T. Hurlbert, E. Linke, G. St Clair & J. Stein. 2006. Scholarly use of information: Graduate students' information seeking behaviour. Information Research, 11. http://informationr.net/ir/11-4/ paper272.html

4. Haglund, L., & P. Olsson. 2008. The impact on university libraries of changes in information behavior among academic researchers: A multiple case study. The Journal of Academic Librarianship. 34: 52–59.

5. Heine, M., I. Winkworth & K. Ray. 2000. Modeling service-seeking behavior in an academic library: A methodology and its application. The Journal of Academic Librarianship. 26: 233–247.

6. Kibrige, H. M., & L. De Palo. 2000. The Internet as a source of academic research information: Findings of two pilot studies. Information Technologies and Libraries: 11–16.

7. Kim, Y., & J. Abbas. 2010. Adoption of library 2.0 functionalities by academic libraries and users: A knowledge management perspective. The Journal of Academic Librarianship. 36: 211–218.

8. Lu, X. B., & J. Guo. 2009. Innovation community: Constructing a new service mode for academic libraries. The Electronic Library. 27: 258–270.

9. Martell, C. 2008. The absent user: Physical use of academic library collections and services continues to decline. The Journal of Academic Librarianship. 34: 400–407.

10. Oakleaf, M. 2010. The value of academic libraries: A comprehensive research review and report. www.acrl.ala.org/value

11. Payette, S. D., & O. Y. Rieger. 1998. Supporting scholarly inquiry: Incorporating users in the design of the digital library. The Journal of Academic Librarianship. 24: 121–129.

12. Pinto, M., V. Fernandez-Marcial, & C. Gomez-Camarero. 2010. The impact of information behavior in academic library service quality: A case study of the science and technology area in Spain. The Journal of Academic Librarianship. 36: 70–78.

13. Ren, W. 2000. Library instruction and college student self-efficacy in electronic information searching. The Journal of Academic Librarianship. 26: 323–328.

14. Shumaker, D., & M. Talley. 2010. Models of embedded librarianship: A research summary. Information Outlook. 14: 27–35.

15. Sidorko, P. E ., & T. T. Yand. 2009. Refocusing for the future: Meeting user expectations in a digital age. Library Management: 30: 6–24.

16. UniSA, 2011. Retrieved 9 June 2011 from www.unisa.edu.au.

17. Weber, R. P. 1990. Basic Content Analysis. Newbury Park, CA: Sage.

CHAPTER 12

Common Knowledge: Learning Spaces in Academic Libraries

Rebecca M. Sullivan

Preus Library, Luther College, Decorah, Iowa, USA

ABSTRACT

The first iterations of the information commons organized library work space around access to digital resources, with research and computing assistance also available. More recently, however, change in academic libraries has paralleled the reorientation of knowledge in higher education. With further involvement in campus-wide initiatives and an emphasis on the social dimension of learning, the learning commons represents transformative change that extends beyond the reach of the traditional academic library. A review of literature on the learning commons provides case studies that reveal collaborative space, partnerships leading to integrated service, and user-centered assessment.

KEYWORDS

Academic libraries, collaboration, learning, learning commons, partnerships

For the past decade, change in academic libraries has paralleled the reorientation of knowledge in higher education. Recently, in line with the emphasis on student-led inquiry and collaborative learning, the learning commons concept has resulted in a trend toward flexible designs and interactive spaces. A review of the literature illustrates quite clearly the development of the learning commons concept. This review was conducted in the Library Literature and Information Science Index and Library Information Science and Technology Abstracts, as well as by following citations to further scholarship.

In the 1990s, the first iterations of the information commons organized work space around access to digital resources, with integrated research and computing assistance (Beagle 1999). At a 2004 conference to celebrate the tenth anniversary of the Information Commons at USC, Beagle presented a matrix that expressed the stages of evolution from information commons to learning commons: adjustment, isolated change, far-reaching change, and transformation (Bailey and Tierney 2008, 2–3). This continuum progressively shifts the emphasis from the facilitation of knowledge discovery in the information commons (imparted through information and technology resources) toward the *creation* of knowledge, enabled by shared learning tasks and productivity tools in the learning commons. Currently, numerous commons are moving further along Beagle's continuum of transformative change toward knowledge-creation activities and integrated support for learning. In Beagle's view, what began "as the reconfiguration of an academic library ultimately becomes a reconfiguration of the learning environment" (2002, 289). Although the terms information commons and learning commons are sometimes used interchangeably in the literature, the focus of this review is on the partnerships and spaces that extend beyond the practice of the information commons; for that reason, the term learning commons will be used unless a specific institution includes information commons in the name of its library space. This focus will point toward spaces that support the heart of higher education: student learning.

1. CAMPUS-WIDE INITIATIVES

One of the central transformative traits of the learning commons is a greater degree of institutional alignment. In a learning commons, libraries are increasing their relevance by extending beyond their own agendas to incorporate campus-wide initiatives. In *Transforming Library Service Through Information Commons: Case Studies for the Digital Age*, Bailey and Tierney portray the extended institutional alignment that distinguishes the learning commons. According to the authors, the learning commons "is clearly and explicitly aligned strategically with the institution-wide vision and mission—that is, as a dynamic and active partner in the broad educational enterprise of the institution, not just the library-centric enterprise" (3). The learning commons' goals are wholly integrated with those of the wider institution. A learning commons does not "simply support but enacts the education mission of the college or university" (Bennett 2008, 184). At this level, the library has greater involvement in a wider range of campus-wide initiatives.

Bennett has been a strong advocate for far-reaching institutional alignment. In his view, the difference between the information commons and the learning commons is "more than a semantic exercise" (184). Clarity about the integration of campus goals will allow a learning commons to design a space that inherently reflects the institutional mission. When it comes to design, Bennett proposes that planners work with initiatives across the campus to "design the commons primarily with the intention that learning will happen there" (2008, 184). This is a view of the library as a learning enterprise more than as an information repository. Unfortunately, Bennett sees a dearth of systematic knowledge about how learning happens in our current libraries. He also decries the fact that self-directed learning "has not inspired library design or propagated a professional literature in the way that digital technology has inspired the information commons" (2005). We seem to persevere on the service operations of libraries rather than planning for their educational impact. Bennett reinforces that libraries need to reorient their planning toward a "systematic knowledge of how students learn"; failing to make that shift will result in libraries that are efficient but not effective (2005). Likewise, Freeman talks about the library as an extension of the classroom, and observes that library facilities are most successful when they are conceived to be an "integral and interdependent part of the institution's total educational experience" (2005).

Toward that end, Bennett provides "First Questions for Designing Higher Education Learning Spaces"—six questions about how a space might be designed to facilitate productive studying, how it might accommodate both solitary and collaborative learning, and how it will enrich educational experiences. Bennett's final word is that planners should endeavor to "understand the learning culture of one's own institution and how it may resonate with and differ from the cultures of other colleges and universities" (2006, 24). Planning for a learning commons is not a one-size-fits-all proposition. Rather than emulating other libraries, such planning involves bringing in campus partners—a broadly based team that includes librarians, administrators, faculty, and students—to engage in the "larger learning and research agendas of the institution" (Stuart 2009b, 18). In fact, in *The Information Commons Handbook*, Beagle suggests that "planning group members should review documentation of every other pertinent campus initiative that offers some potential relationship to LC services and resources" (2006, 84).

While many planning documents for learning commons exist on the Web, Victoria University in Melbourne, Australia, provides a strong model of a planning process that was enacted in response to the wider institutional mission. VU's stated mission is to "transform the lives of individuals and develop the capacities of industry" in the western

Melbourne region. Keating, Kent, and McLennan identify the learner-centered approaches that are embedded in VU's teaching and learning policies, and demonstrate how the learning commons' integration of services and functions support the university-wide mission (2008, 300). A broadly based planning group, including the University Librarian, the Chancellor of Education Services, the Chancellor of Teaching and Learning Support, and other campus administrators, was brought together to develop a space that responds to VU's emphasis on learner-centered practice. The new learning commons offers physical and virtual resources that are conducive to students becoming more proactive and autonomous, but the space also provides integrated support services nearby. Having designed a space that successfully reflects VU's mission, the authors emphasize the point that the learning commons would have minimal impact if it were not part of a campus-wide approach (Keating and McLennan 2005).

From another angle, a collaborative campus effort for the learning commons can also crystallize wider institutional goals. When the Kate Edger Information Commons at the University of Auckland (New Zealand) created an integrated environment for learning support, it actually acted as a catalyst for a comprehensive e-Literacy program. Mountifield documents that prior to the development of the facility, the university published a recommendation to develop "a coherent approach to training programmes in computer and information literacy skills," but there was little progress in implementation. However, with the collocation of the Student Learning Centre, The English Language Self Access Centre, and an integrated technology and information literacy Help Service in the new building, the commons facilitated the needed collaboration to put plans into action. In Mountifield's view, the commons promotes new partnerships and "provides the infrastructure for the functional integration" of services (2004, 83–84).

2. PARTNERSHIPS

In order to enact an institutional mission, a learning commons must reach across campus to form partnerships with other initiatives. In "New Library Facilities: Opportunities for Collaboration," Lippincott notes that most library-affiliated collaborations take place in renovations or additions. Being alert to the possibility of new partnerships offers many potential benefits, namely, opportunities "to provide seamless services to users, to leverage the various talents that different professional groups can bring to a service, and to pool institutional resources" (2004, 150). Using the Edger facility as a case study, Beatty and Mountifield outline the advantages of integrated learning services and review the qualities of successful partnerships, including strong initial and ongoing planning, clear documentation and communication, mutual commitment and trust. The foundation of a strong partnership lies in the articulation of a shared vision and objectives in the context of the institutional climate (2006).

Recognizing various barriers to collaboration, Lippincott agrees that a statement of partnership can help to articulate a common vision for ongoing, shared planning. She also points to exemplars where "aligning the mission of a collaborative facility to the overall institutional mission" provides a solid framework to build integrated services (2004, 150). Barratt and Daniels' environmental scan suggests that "coming to an agreement on the overarching mission and identity" of common spaces can be a challenge (2008, 9). Cooperating units have to concentrate on their common goals rather than on their differences. However, Fister observes the increase in nonlibrary units in library buildings and notes that collaborative partners do not need to be "administratively merged to work together." She highlights the importance of units being tied "not by reporting lines, but rather through the intersections of their missions" (2004, 3).

2. Partnerships

While the rewards of collaboration are extolled by Wilson in the ARL report "Collaborate or Die: Designing Library Space," there are also challenges. Based on lessons learned at the University of Washington, she reports that strong collaborations have essential elements, including the commitment of organizations, formal structures for division of labor and levels of communication, dispersed authority to balance ownership, shared resources as well as rewards, and "mutual respect, trust, mentoring, and humor." Revealing the complexity of collaboration, she observes that

Collaboration is a choice. It can't be mandated. It's hard work. It's fragile. Collaborative design doesn't come naturally [...] Collaborative design requires negotiating skills, making tradeoffs, and sharing control. Collaborators learn how to cross boundaries and have a high tolerance for ambiguity. Successful collaborators get beyond the subtle barriers created by their professional roles [...] Innovative organizations pay attention to supporting the skills and providing the latitude needed in collaborative design. (2002, 3–4)

That being said, Wilson uses the UWired pilot at the University of Washington to demonstrate how collaborative design can be its own reward. Following from the provost's initiative to "do something about technology in learning," a team from the library, computing, and undergraduate education worked together to create a "collaboratory," a networked classroom for group learning. UWired has since grown to realize more collaboratories, a faculty development center, a center for multimedia production, a digital animation lab, and a program on educational technology. Wilson argues that the UWired experience is enriched by the "symbiotic relationship between traditional library functions and 'nonlibrary' functions" (3).

A vibrant collaboration benefits from the sense of newness that accompanies a shared challenge. In another account of the University of Washington partnerships, McKinstry states that partners "tend to be more open to diverse solutions when they realize that the problem at hand is more complex," and she observes that the best partners thrive on innovation. McKinstry highlights the social and intellectual benefits of the Odegaard Undergraduate Library's "Research Exposed" lecture series, the program of exhibits and displays, and the undergraduate research award—all partnerships with other campus/community interests—and concludes, "libraries make good partners." While the library is generally seen as a neutral party, its service orientation, management skills, and interdisciplinary nature make it well-situated for collaboration (2004, 141).

The ongoing partnership that both Wilson and McKinistry describe has become a model for managing the campus-wide conversation that leads to a learning commons. Beagle uses UWired as a model for how "the campus conversation becomes the seedbed for a collaborative process that can begin within the library and extend outward, or can be the linchpin in a process already initiated by university leadership as part of a broader campus agenda" (2006, 108). UWired serves as an early demonstration of how the library can be involved in far-reaching change.

The literature reflects the range of innovative partnerships taking place among libraries, writing and academic support centers, teaching and learning centers, disability coordinators, diversity centers, service learning initiatives, undergraduate advising programs, and digital centers. Loyola University was one of the first to open a facility for academic support services in the library. The Academic and Career Excellence Center—or ACE Center—brings tutoring, academic assessment, disability services, counseling, career guidance, and research help to the Monroe Library, with peer tutors who make referrals to the various services (Orgeron 2001). Another engaging partnership exists at the University of Guelph in Ontario. Kaufman and Schmidt describe how the learning commons in McLaughlin Library offers an integrated program of learning, writing, ESL, research, technology, and disability services. Based on a partnership among Learning and Writing Services, the library, Computing and Communications Services, and Teaching Support Services, the Guelph Learning Commons combines staff expertise and a research-based peer educator program to provide these comprehensive services. Moreover, the strongly theoretical design of services has positioned the learning commons staff to serve as a campus resource for pedagogy, involving the commons "in a number of campus-wide initiatives that impact on student learning" (2005, 253).

Alongside other academic support units, another common partner in integrated services is the writing center. The process-oriented approach that is shared by writing tutors and reference librarians hints at the affinity between these two services. Bledsoe and Cooke identify challenges common to librarians and writing centers—guiding students through the process when they face deadlines and uncertainty about assignment guidelines, and teaching students to use sources wisely—and share how colleagues at Florida Gulf Coast University seek collaborative solutions (2008). Mahaffy documents two pilot projects at New Mexico State University to affirm a principle found elsewhere in the literature: the most successful integrated research/writing support services seem to be housed within the library. In the close proximity of the learning commons, librarians and writing consultants become more familiar with the other's expertise, communicate more readily, and easily share referrals (2007).

One of Mahaffy's case studies is the writing center in the Giovale Library of Westminster College, a partnership that is also discussed in Malenfant's "The Information Commons as a Collaborative Workspace." Malenfant reports that at Westminster, the writing center is run by an English professor who, with his staff, contributes to the design of library space and participates fully in commons activities. In fact, at Westminster, leaders from circulation, the lab, the writing center, and reference have a shared investment in the cross-training for IC tutors, which leads to "greater understanding among all IC staff of the various resources and services offered by its various branches" (2006, 284). Fliss touts a similar model in Dartmouth College's Research, Writing, and Information Technology Center (RWIT). Writing about a number of partnerships in the Baker-Berry Library, Fliss states that student support needs "can no longer be met by a single academic department or student service" (2005, 378). In this spirit, RWIT provides peer support in composition, research, and technology. Evidence of the success of the RWIT can be seen in its full schedule and the expressed interest of other campus partners to join this collaboration (Stuart 2009a, 13).

The University of Massachusetts at Amherst's W.E.B. Du Bois Library provides yet another example of a productive library/writing center partnership. As part of a study investigating user preferences, Moore and Wells report that students still look for face-to-face assistance and embrace the integrated academic services of the learning commons. One student in the study remarks that the writing center has been a "prime contributor of academic success." Another asks for extended hours. Further testimony exists in the 33% increase in tutoring sessions in the first semester after the writing center moved to the learning commons (2009, 77). The potential of writing center partnerships can be seen in an ALA collection of twelve essays,*Centers for Learning: Writing Centers and Libraries in Collaboration.* In addition to exploring the theoretical foundation of libraries and writing centers, the authors present case studies, compile guidelines for sustainable partnerships, and outline tutor cross-training programs (Elmborg and Hook 2005).

In support of integrated programs, Beatty and White have conducted research to demonstrate that collaborative service models are more effective. After conducting an environmental scan of North American information commons, they matched the level of collaboration in the commons with six elements that are linked to learning support. The researchers found that "the more collaborative the commons, the more likely it is to offer design and service features which enable integrated e-literacy learning" (2005, 8). Students benefit from services that are available in close proximity, where the various types of expert assistance are delivered in coordination.

Once integrated services are assembled in the learning commons, managing student access becomes an important consideration. The Saltire Centre at Glasgow Caledonian University in Scotland is a good representation of one approach to mediating services. After reviewing student access to services, the Saltire Centre developed The Base, a support desk that answers basic questions and makes referrals to specialist services including "library services, IT systems (virtual learning environment, etc.), subject librarians, registry, careers, effective learning service, funding, well-being advice, international student support, student disability service, research collections, counseling, nursery and chaplains." Located in the midst of a group study area, The Base provides the coordination for a supportive environment and is supplemented by Base staff roving throughout the building. Following referrals, most specialists meet with students in shared consulting rooms, bringing their services a step closer to the social area of the commons (Howden 2008, 213).

In addition to academic and writing support, the range of services in many learning commons includes services for digital multimedia. For example, along with tutoring, the University of Tennessee's Hodges Library offers The Studio and a Digital Media Service. Dewey has written about these projects as part of the circle of service philosophy that drives the learning commons at the university. She recounts the phases of transformation that led to The Studio—a self-production facility with experts on hand, where students and faculty can "combine digital media in creative ways" for assignments and scholarly resources (2008, 92). Next to The Studio, The Digital Media Service is a drop-off production facility where staff "digitize and store instructional materials for faculty from all Colleges, with the ability to convert a variety of formats into digital media." These collaborative services, directed by an interdisciplinary advisory committee, have become central to "course-related digital production" at UTK (Dewey 2002, 1–2). Similarly, at Georgia Tech, students improve their multimedia skills using the popular software in the productivity center of the West Commons. Supported by a nearby help desk and presentation practice space, the commons has twenty-five workstations equipped with high-end multimedia software to support coursework. After a year of operation, the Multimedia Studio set up "short courses in multimedia software," and faculty began to bring entire classes to the facility for training (Stuart 2008b, 338). Multimedia production is just one of the services to support learning in the Georgia Tech Commons, along with undergraduate advising, a tutoring program, counseling seminars, and a Center for Teaching and Learning.

Together with the services of the learning commons, libraries are increasingly developing virtual commons to complement the physical learning space. The virtual commons offers access to electronic resources and social networking tools, resources that encourage exploration and discussion beyond the walls of the library. Lippincott locates many of the goals of the virtual commons in the University of Southern California's Leavey Library's efforts to understand the needs of today's learners. At the University of Southern California and elsewhere, Net Generation students have mobile technology that allows them to multitask anywhere, anytime—creating a preference for self-service and interactivity on the Web. Seeking user-centered delivery, the virtual commons at Leavey integrates library pathfinders and tutorials into the course management system, bringing resources to where students will be engaged in their studies. Employing Web 2.0 tools, blogs allow classmates to communicate about courses, and easy chat services put students in touch with librarians when needed. Responding to yet another characteristic of Net Gen learners, Leavey's Web services rely on graphic design to appeal to the visual orientation of students. A student team informs the design of virtual services to "ensure that new services will be responsive to both their needs and their style" (2005, 58).

The University of Indiana, Bloomington also supports its thriving physical learning environment with virtual outreach. Using the course management system, librarians create "class pages" for research assignments, answer e-mail and chat questions, and provide online tutorials (Dallis and Walters 2006). The University of Manitoba has received attention for its virtual commons as well. Created to support an international students project, the "Virtual Learning Commons" Website provides information literacy tutorials, an assignment manager, an online writing tutor, academic support resources, and social networking tools to allow students to interact outside of the classroom. Making these online resources available to students extends the reach of the library into their environments, into their spheres of learning (http://www.umanitoba.ca/virtuallearningcommons/).

In addition to supporting students, learning commons are increasingly partnering to support faculty and enhance instruction. At the University of Calgary in Alberta, the library and IT work with the Teaching and Learning Centre to help faculty learn technical skills and best practices for teaching with technology. Crossing organizational boundaries, these units work together to assist faculty with the learning management system, new applications, and emerging technologies. One of the partnership's most successful ventures has been a series of workshops, now an annual event, called "Faculty Technology Days," where topics range from "an introduction to a new database, to new social software, to a panel discussion on plagiarism" (Beatty and Mountifield 2006, 241). For the commons staff, these early collaborative efforts have led to expanded relationships with faculty and new avenues to support learning.

While productive relationships often lead to further endeavors, such as at Calgary, partnerships can also be tenuous. When the Johnson Center at George Mason University opened in 1995 (housing a library, a student activity center, an academic resource center, and a computer center) it was conceived of as a "focal space" for students. An interdisciplinary undergraduate program called the New Century College also moved into the building, bringing important connections to the Freshman Center, the Center for Teaching Excellence, and the Writing Center. Unfortunately, when the New Century College was subsumed into the College of Arts and Sciences, it was relocated. In retrospect, Gibson and Lockaby write, "The implicit learning commons concept behind the Johnson Center Library has remained only partly realized […] programmatic partnerships between the Library and other building units have not always sustained themselves in a way that would show deep impact." For a number of reasons, the mission of the Johnson Center Library is currently under review, leaving "powerful lessons about the need for deep collaboration, thoughtful planning, and ongoing experimentation" (2007, 329). Placing units in proximity doesn't necessarily ensure thriving partnerships. Relationships must be forged early in the planning process and participants must be committed to ongoing experimentation and repurposing.

3. LEARNING IN THE COMMONS

In the unfortunate scenario described above, all is not lost. Aside from the current uncertainty of the Johnson Center's identity, part of the library's original concept survives because of the vibrant social element in this multipurpose building. Gibson and Lockaby take heart that students still find the library to be a conducive environment for studying, working together, and socializing. The authors observe that "the unplanned and fortuitous learning that occurs in the building as a whole is palpable and powerful" (2007, 328). Many learning commons host cultural events and exhibits to engage students in that sort of fortuitous learning. In the Gould Library Athenaeum, an "elegant" reading room at Carleton College, librarians join with faculty to host "poetry readings and author events, debates, concerts, discussion forums, and lectures." Demas

3. Learning in the Commons

compares this range of activity to the role of the Mouseion at Alexandria in Ancient Greece—home to music, poetry, a gallery, a library, and an assemblage of scholars. Demas sees libraries as intellectual and social commons that "engage the community in discourse and in enjoyment of the life of the mind" (2005, 34). The University of Michigan also highlights the practice of building community with a gallery in the new Harlan Hatcher Graduate Library. In collaboration with campus and community partners, the library mounts student exhibits, hosts lectures, and invites nonlibrary programming (Stuart 2009a, 31). Similarly, in an effort to create a campus hub, the Saltire Centre Learning Commons in Glasgow hosts many campus events, from a fashion show catwalk to conventional exhibits. On behalf of the Saltire Centre, the Associate Director in Learning Support explains, "All of this activity comes by basically saying yes to most requests and then working out the logistics" (Howden 2008, 211). The serendipitous learning that takes place at such events, in the words of Beagle, promotes "social inclusion in a college or university as members of a community of learning" (2006, 35).

There is a kind of learning that happens outside the classroom, in the spontaneous and informal interaction of campus life. Herein lies the capacity of the learning commons to bring about the social dimension of learning. The 2006 Joint Information Systems Committee report, "Designing Spaces for Learning," pointed to the social mechanism of the Saltire Centre when it reported that as the "social heart the of campus, [it is] a place where students meet and converse as well as study." In observing how the Saltire Centre capitalizes on the social environment, a report from the Scottish Funding Council comments that "it goes further by making itself the starting point of the learning process and by encouraging 'deliberate socializing.' This includes accepting noise [...] The Saltire Centre could be seen as an unstructured 'educational soup'" (2006, 45–46). Obviously, higher education has endeavored to exploit the social basis of learning. In "Libraries Designed for Learning," Bennett recalls how faculty have adopted active learning strategies, and have "built experiential and problem solving materials into their courses [...] to engage with the social dimensions of learning and knowledge" (2003, 3). This discussion necessarily leads to Bruffee's concept of knowledge as a community project, which Bennett alludes to when he asserts that a space which "celebrates the communal character of knowledge will indeed foster learning" (2005). Constructivist learning theory shifts the locus of knowledge from the professor to the collective discovery of students. Individuals make meanings through experience and interaction with others. Learning is now an active process that emphasizes the need for collaboration. Aided by the provision of electronic resources to facilitate a self-directed learning environment, the nature of the learning commons accords with the way that social constructivism has changed higher education.

4. DESIGNING SPACES FOR LEARNING

In the spirit of collaborative learning, spaces for group work have become a hallmark of the learning commons. Bennett states that "library directors reported providing group study space much more frequently than one would expect" (2003, 17). He provides selected excerpts from the survey results that are representative of the literature regarding group study:

Just the most notable thing about usage is [...] the extreme growth in group study [...] We're seeing that virtually all of [the tables] are filled with students working together [...] This space will be filled, literally every chair [...] and they're all talking at the same time. And the hum that rises above this is just amazing. And they don't care [...] There's all this din that occurs [from] hundreds of students in this same space, all working together and all talking at the same time [...] and they're all clustered around the computers as well, working together in some cases. (17)

Naturally, the growth of group study—a reflection of current pedagogy—is apparent in most learning commons case studies. Sharing the rationale behind the University of Akron's planning for a learning commons, Franks (2008) demonstrates the value of workstations to accommodate group research. Shill and Tonner's 2003 study on physical improvements in academic libraries shows that demand for group spaces tends to exceed availability. Likewise, Gardner and Eng's survey (2005) of Net Generation students at the University of Southern California confirms that many students use the Leavey Library to study with a group, and more than half of respondents want more collaborative workrooms.

The learning commons is clearly designed to meet the diverse needs of learners, whether in groups or as individuals. Looking more closely, the Gardner and Eng study also shows that eighty percent of respondents in Leavey Library report coming to study alone, indicating that the need for quiet study areas remains (2005, 412). Dallis and Walters report that at Indiana University, Bloomington, "students request quiet, reflective space in the library as often as they request more technology" (2006, 258). Barratt and Daniels' environmental scan of the learning commons model reveals a general increase in traffic in libraries; however, "many of the respondents reported that noise is one of the biggest challenges […] higher noise levels hinder students' concentration on academic work and disturb librarians in consultation" (2008, 8).

Taken a step further, Gayton (2008) questions the "social" model of the academic library, making a thoughtful assertion of the need for quiet, individual study spaces. Gayton has scoured library literature for a persuasive set of arguments advancing the merit of "communal," yet independent, studious work stations, suggesting that there are effective ways to mediate the incompatibility of quiet and social spaces. Interestingly enough, in the results of an ARL study on innovation in libraries, there is an entry from the University of Tennessee that claims that UTK is "in the process of designing more traditional quiet / reading room space" (Stuart 2009a, 47). Many libraries have been challenged by trying to meet a wide variety of user needs.

Sinclair sees the answer to diverse student needs in a reinvention of library spaces, and he promotes adaptation to new ways of learning. With the freedom of wireless technology, he advocates "flexible workspace clusters that promote interaction and collaboration, and comfortable furnishings [… to] encourage creativity and support peer-learning" (2007, 4). His principles for design include variety, flexibility, and mobility. In Sinclair's vision, the Commons 2.0 has tables with "organic shapes—kidney, oval, half-circle—that encourage inclusiveness and participation. Some tables are movable, allowing different group sizes and configurations" (5). To meet the diverse needs of learners, there are sofas, soft chairs, and task chairs—best when on wheels—among the furnishings. Allowing students to shape the environment gives them more control of their learning. There is a full array of applications, digital resources, and productivity software on each workstation in the Commons 2.0, some equipped with widescreen monitors where a few students can gather, and integrated services are nearby to provide expert assistance.

There is yet another resource that sustains students and nourishes learning: food. Survey results from Bennett's 2003 Council on Library and Information Resources study reveal that, in the nearly 250 responding libraries that were built or renovated in the years 1992–2001, "50% of the projects included vending machine food and beverages, while 23% reported including staffed food services and another 27% reported some other type of food service" (18). Bennett concludes that the tide has turned; new or renovated library space generally provides some kind of food service. Bennett (2005) compellingly makes the case that food domesticates a space, fostering the kind of informal, serendipitous conversation that leads to learning. Many learning commons have tried to leverage this kind of fruitful interaction among students and faculty by installing a coffee bar or café on the premises.

The availability of sustenance also allows students to stay in the library longer, increasing time on task. The Georgia Tech library reports that its café has increased library usage: "at certain times of the day, the café bristles with laptops and flipcharts that are dragged in from nearby study areas, and serves as a communing ground for students and faculty" (Stuart 2009b).

The learning commons comes with a trend toward more human-centered design. How are planners to know which elements will humanize a space? The answer, according to Brown and Long, would be to ask users. They support the notion that "learning environments should be developed by those who will use them" (2006, 9.5). In fact, there is a voice throughout the literature on designing learning commons that says, "Why not give them what they want?" Like the classroom, when we renegotiate authority in the library, student ownership for learning increases.

Following from that logic, there are several good models for user-centered design in the literature. Collins and Somerville describe students using Web 2.0 tools to co-design elements of libraries at San Jose State and California Polytechnic State Universities. In addition to giving librarians and stakeholders insight into student needs, a co-design team creates ongoing relationships that lead to an iterative design process. The authors contend that the approach moves beyond "library centric" thinking to focus on "what's best for users" (2008, 808).

Development of the learning commons requires that libraries assess and customize learning spaces to their users. Fox and Stuart show how Georgia Tech used student focus groups, field studies, and the Student Advisory Council to design the renovation of the Library East Commons. The authors discovered that their "users knew more than anyone else about the amenities and qualities of good learning spaces" (2009). Fortunately, Stuart captures many of the discovery techniques that were used—surveys, assessment documents, process highlights, traps to avoid, tips for advisory groups, and instructions for design charrettes—and publishes them on the Web. The space planning toolkit, along with other resources, can be found on the "Transforming Research Libraries" Website of the Association of Research Libraries (Stuart 2008a). Educause also has a Web version of a "Student Input on Learning Spaces Tool," an activity that helps colleges and universities see their campuses as their students do (Educause Learning Initiative 2006).

Both the ARL and the Educause tools use anthropological field techniques advanced by Foster and Gibbons' *Studying Students: The Undergraduate Research Project at the University of Rochester*. In this influential study, the authors set out to answer the question, "What do students *really* do when they write research papers?" The purpose of the study is to understand how students work. The authors state, "Once we understand this, we set about to support the work practices that will help our students, and the library and the university, succeed. This, for us, is user-centered design" (2007, 82). The team uses ethnographic interviews, flipcharts in public areas, and charrette-style workshops to collect information, turning their insights into new design ideas.

5. CHANGE

All of the above design techniques emphasize ongoing assessment. Baer, Belliston, and Whitchurch reinforce the ongoing nature of evaluation in their account of the early stages of the commons at Brigham Young University. From the start, planners saw an evaluation program as an essential part of the development of the commons (2006). Although Whitchurch admits that the commons community still struggles with the assessment issue, he presents a systematic study that confirms group use patterns in the Harold B. Lee Library. After two years of operation, design features were changed in response to the study's data, and, perhaps as importantly, the library now has a basis for longitudinal comparison to gauge future changes (2009). Commons are by nature "works in progress." The director of the new learning commons at North Carolina State University includes ongoing assessment among his primary responsibilities: "I'm out there trying to make connections […] trying to identify what our users need. And that outreach will be an ongoing effort, even long after the Learning Commons opens." Williams compares the learning commons project to a workshop—an ongoing seminar—where staff learn "how our users like to work and how we can support collaboration" (Spencer 2007, 311).

The remark is made frequently in the literature that there is little standardized assessment of how learning happens in libraries. Yet, there are sources that address a systematic plan to measure outcomes of the learning commons. Both Beagle's *The Information Commons Handbook* and Bailey and Tierney's *Transforming Library Service Through Information Commons: Case Studies for the Digital Age* have informative chapters that differentiate various types of assessment—in fact, Beagle's book includes surveys from the universities of Southern California, Arizona, and Calgary—but there is no standard tool specifically tailored to the learning commons. MacWhinnie conjectures, "Perhaps an impediment to assessment in general is the inability to evaluate the multiple features of this new learning resource." The learning commons is a new model that extends "beyond the scope of the traditional library and will therefore require new methods of assessment to determine its effectiveness" (2003, 251). The additional challenge, of course, is that every learning commons is different, calling for a customized assessment tool to measure outcomes against the goals of the institution and the needs of its students.

In support of diverse campus initiatives, involving a variety of partnerships, the single unifying element of the many manifestations of the learning commons is *change*. Staged evolution is a core principle in Beagle's conception of the continuum from information commons to learning commons; evolution is also inherent in the development of each learning space. Ongoing assessment and the scholarly literature surrounding the commons inform the process. Relying on the shared knowledge and experience of our colleagues, librarians and their partners continue to reshape library space to best foster student learning, make an appraisal of the outcome, and refine yet another iteration of user-centered space.

REFERENCES

1. Baer, W., Belliston, C. J. and Whitchurch, M. J. 2006. Information commons at Brigham Young University: Past, present, and future. *Reference Services Review*, 34: 261–78.
2. Bailey, D. R. and Tierney, B. G. 2008. *Transforming library service through information commons: Case studies for the digital age*, Chicago: American Library Association.
3. Barratt, C. C. and Daniels, T. 2008. "What is common about learning commons? A look at the reference desk in this changing environment". In *The desk and beyond: Next generation reference services*, Edited by: Steiner, S. K. and Madden, M. L. 1–12. Chicago: American Library Association.
4. Beagle, D. 1999. Conceptualizing an Information Commons. *The Journal of Academic Librarianship*, 25: 82–89.
5. Beagle, D. 2002. Extending the information commons: From instructional testbed to Internet2. *The Journal of Academic Librarianship*, 28: 287–96.
6. Beagle, D. 2006. *The information commons handbook*, New York: Neal-Schuman.
7. Beatty, S. and Mountifield, H. Collaboration in an information commons: Key elements for successful support of e-literacy.

Leicestershire, UK: Paper presented at the 5th International Conference on e-Literacy.https://dspace. ucalgary.ca/ bitstream/ 1880/44776/1/beatty-mountifield.pdf

8. Beatty, S. and White, P. 2005. Information commons: Models for eLiteracy and the integration of learning. *Journal of eLiteracy*, 2: 2–14.

9. Bennett, S. 2003. "Libraries designed for learning". Washington, DC: Council on Library and Information Resources.http://www. clir. org/pubs/reports/pub122/pub122web.pdf

10. Bennett, S. 2005. "Righting the balance". In *Library as place: Rethinking roles, rethinking space*, Washington, DC: Council on Library and Information Resources. http:// www.clir. org/ pubs/ reports/pub129/bennett.html

11. Bennett, S. 2006. First questions for designing higher education learning spaces. *The Journal of Academic Librarianship*, 33: 14–26.

12. Bennett, S. 2008. The information or the learning commons: Which will we have?. *The Journal of Academic Librarianship*, 34: 183–85.

13. Bledsoe, C. and Cooke, R. 2008. Writing centers and libraries: One-stop shopping for better term papers. *The Reference Librarian*, 49: 119–28.

14. Brown, M. and Long, P. 2006. "Trends in learning space design". In *Learning Spaces*, Edited by: Oblinger, D. G. 9.1–9.11. Washington, DC: Educause. http://www.educause. edu/ir/library/ pdf/ PUB7102i.pdf

15. Colins, L. and Somerville, M. M. 2008. Collaborative design: A learner-centered library planning approach. *The Electronic Library*, 26: 803–20.

16. Dallis, D. and Walters, C. 2006. Reference services in the commons environment. *Reference Services Review*, 34: 248–60.

17. Demas, S. 2005. "From the ashes of Alexandria: What's happening in the college library?". In *Library as place: Rethinking roles, rethinking space*, Washington, DC: Council on Library and Information Resources.http://www.clir.org/pubs/reports/pub129/demas.html

18. Dewey, B. 2002. "University of Tennesee's collaborative digital media spaces". In *ARL Bimonthly Report*http://www.arl. org/bm~doc/collabtenn.pdf

19. Dewey, B. 2008. Social, intellectual, and cultural spaces: Creating compelling library environments for the digital age.*Journal of Library Administration*, 48: 85–94.

20. Educause Learning Initiative. 2006. "ELI discovery tool: Student input on learning spaces tool".http://www.educause.edu/ ELI/ELIDiscoveryToolStudentInputon/156829

21. Elmborg, J. K. and Hook, S. 2005. *Centers for learning: Writing centers and libraries in collaboration*, Chicago: Association of College and Research Libraries.

22. Fister, B. 2004. Common ground: Libraries and learning. *Library Issues*, 25: 1–4.

23. Fliss, S. 2005. Collaborative creativity: Supporting teaching and learning on campus. *College and Research Libraries News*, 66: 378–407. http://ala.org/ala/mgrps/divs/acrl/publications/crlnews/2005/may/colcreativity.cfm

24. Foster, N. F. and Gibbons, S. 2007. *Studying students: The undergraduate research project at the University of Rochester*, Chicago: Association of College and Research Libraries.

25. Fox, R. and Stuart, C. 2009. Creating learning spaces through collaboration: How one library refined its approach. *Educause Quarterly*, : 32 http://www.educause.edu/EDUCAUSE± Quarterly/EDUCAUSEQuarterlyMagazineVolum/CreatingLearningSpacesThroughC/163850

26. Franks, J. A. 2008. Introducing learning commons functionality into a traditional reference setting. *Electronic Journal of Academic and Special Librarianship*, : 9 http://southernlibrarianship. icaap.org/content/v09n02/franks_j01.html

27. Freeman, G. T. 2005. "The library as place: Changes in learning patterns, collections, technology, and use". In *Library as place: Rethinking roles, rethinking space*, Washington, DC: Council on Library and Information Resources.http://www.clir. org/pubs/reports/pub129/freeman.html

28. Gardner, S. and Eng, S. 2005. What students want: Generation Y and the changing function of the academic library. *Portal: Libraries and the Academy*, 5: 405–20. [CrossRef], [Web of Science ®]

29. Gayton, J. T. 2008. Academic libraries: "Social" or "communal?" The nature and future of academic libraries. *The Journal of Academic Librarianship*, 34: 60–6. [CrossRef], [Web of Science ®]

30. Gibson, C. and Lockaby, D. C. 2007. The Johnson Center Library at George Mason University. *Reference Services Review*, 35: 322–30.

31. Howden, J. 2008. "The Saltire Centre and the learning commons concept". In *Learning commons: Evolution and collaborative essentials*, Edited by: Schader, B. 201–26. Oxford, UK: Chandos Publishing.

32. Joint Information Systems Committee. 2006. "Designing spaces for effective learning: A guide to 21st century learning space design".

Birmingham, UK: Publication from the JISC Conference. March 14. http://www.jisc.ac.uk/eli_learningspaces.html

33. Kaufman, J. and Schmidt, N. 2005. Learning commons: Bridging the academic and student affairs divide to enhance learning across campus. *Research Strategies*, 20: 242–56.

34. Keating, S., Kent, P. G. and McLennan, B. 2008. "Putting learners at the centre: The learning commons journey at Victoria University". In *Learning commons: Evolution and collaborative essentials*, Edited by: Schader, B. 297–324. Oxford, UK: Chandos Publishing.

35. Keating, S. and McLennan, B. 2005. "Making the links to student learning. Paper for Academic and Vocational Education Boards, Victoria University, Melbourne, Australia". http://74. 125.155.132/scholar?q=cache:jzSrdjRbftUJ:scholar.google.com/&hl=en&as_sdt=2 000

36. Lippincott, J. K. 2004. New library facilities: Opportunities for collaboration. *Resource Sharing and Information Networks*, 17: 147–57.

37. Lippincott, J. K. 2005. Net generation students and libraries. *Educause Review*, : 40http://www.educause.edu/ir/library/pdf/erm0523.pdf

38. MacWhinnie, L. A. 2003. The information commons: The academic library of the future. *Portal: Libraries and the Academy*, 3: 241–57.

39. McKinstry, J. 2004. Collaborating to create the right space for the right time. *Resource Sharing and Information Networks*, 17: 137–46.

40. Mahaffy, M. 2007. Exploring common ground: U.S. writing center/library collaboration. *New Library World*, 109: 173–81.

41. Malenfant, C. 2006. The information commons as a collaborative workspace. *Reference Services Review*, 34: 279–86.

REFERENCES

42. Moore, A. C. and Wells, K. A. 2009. Connecting 24/5 to Millennials: Providing academic support services from a learning commons. *The Journal of Academic Librarianship*, 35: 75–85.

43. Mountifield, H. 2004. The Kate Edger information commons—a student-centred learning environment and catalyst for integrated learning support and e-Literacy development. *Journal of eLiteracy*, 1: 82–96.http://researchspace.auckland.ac.nz/handle/2292/435

44. Orgeron, E. 2001. Integrated academic student support services at Loyola University: The library as a resource clearinghouse. *Journal of Southern Academic and Special Librarianship*, : 2http://southernlibrarianship.icaap.org/content/v02n03/orgeron_e01.htm

45. Scottish Funding Council. 2006. "Spaces for learning in further and higher education. Report for the Scottish Funding Council". Edinburgh: AMA Alexi Marmot Associates. http:// www. jiscinfonet. ac.uk/Resources/external-resources/sfc-spaces-for-learning

46. Shill, H. B. and Tonner, S. 2003. Creating a better place: Physical improvements in academic libraries, 1995–2002. *College and Research Libraries*, 64: 431–66. [Web of Science ®]

47. Sinclair, B. 2007. Commons 2.0: Library spaces designed for collaborative learning. *Educause Quarterly*, 4: 4–6.http:// www. educause.edu/EDUCAUSE±Quarterly/EDUCAUSEQuarterlyMagazineVolum/Commons20LibrarySpacesDesigned/162265

48. Spencer, M. E. 2007. The state-of-the-art: NCSU libraries learning commons. *Reference Services Review*, 35: 310–21.

49. Stuart, C. 2008a. "ARL learning space pre-programming toolkit". Association of Research Libraries.http://www.arl. org/bm~ doc/planning-a-learning-space-tool-kit.pdf

50. Stuart, C. 2008b. "Improving student life, learning and support through collaboration, integration, and innovation". In*Learning*

commons: Evolution and collaborative essentials, Edited by: Schader, B. 325–58. Oxford, UK: Chandos Publishing.

51. Stuart, C. 2009a. "Innovative spaces in ARL libraries: Results of a 2008 study". Association of Research Libraries.http:// www.arl. org/rtl/space/2008study/

52. Stuart, C. 2009b. Learning and research spaces in ARL libraries: Snapshots of installations and experiments. *Research Library Issues*, 264: 7–18. http://www.arl.org/bm~doc/rli-264-spaces.pdf

53. Whitchurch, M. J. 2009. Evaluating group use of the information commons. *College & Undergraduate Libraries*, 16: 71–82.

54. Wilson, L. 2002. Collaborate or die: Designing library space. *ARL Bimonthly Report*, www.arl.org/bm~doc/collabwash.pdf

55. University of Manitoba. "Virtual learning commons". http://www.umanitoba.ca/virtuallearningcommons/

ADDITIONAL READINGS.

1. Arp, L. and Woodard, B. S. 2005. Beyond classroom construction and design: Formulating a vision for learning spaces in libraries. *Reference and User Services Quarterly*, 44: 296–300.
2. Bennett, S. 2008. Libraries and learning: A history of paradigm change. *Portal: Libraries and the Academy*, 9: 181–97.
3. Brown, M. 2005. "Learning spaces". In *Educating the net generation*, Edited by: Oblinger, D. G. and Oblinger, J. L. Washington, DC: Educause. http://www.educause.edu/educatingthenetgen
4. Church, J. 2005. The evolving information commons. *Library Hi Tech*, 23: 75–81.
5. Fitzpatrick, E. B., Lang, B. W. and Moore, A. C. 2008. Reference librarians at the reference desk in a learning commons: A mixed

methods evaluation. *The Journal of Academic Librarianship*, 34: 231–38.
6. Gabb, R. and Keating, S. 2005. "Putting learning into the learning commons: A literature review". Melbourne, , Australia: Working Paper, Victoria University. http://eprints.vu.edu.au/94/
7. Haas, L. and Robertson, J. 2004. *SPEC Kit 281: The information commons*, Washington, DC: Association of Research Libraries.
8. Held, T. 2008. The information and learning commons: A selective guide to sources. *Reference Services Review*, 37: 190–206.
9. Holley, R. and Steiner, H. 2009. The past, present, and possibilities of commons in the academic library. *The Reference Librarian*, 50: 309–32.
10. Misenick, K. E., O'Connor, J. S. and Young, J. 2005. Ten years in the life of a new kind of campus center. *About Campus*, 10(3): 8–16.
11. Roberts, R. L. 2007. The evolving landscape of the learning commons. *Library Review*, 56: 803–10.
12. Schader, B. 2008. *Learning commons: Evolution and collaborative essentials*, Oxford, UK: Chandos Publishing.
13. Sinclair, B. and Collaborative learning commons. "Bibliography and links".http://sites.google.com/site/collaborativelearningspaces/home/bib-links
14. Sinclair, B. and Collaborative learning commons. 2009. The blended librarian in the learning commons. *College and Research Libraries News*, 70: 504–7.
15. Spencer, M. E. 2006. Evolving a new model: The information commons. *Reference Services Review*, 34: 242–47.

CHAPTER 13

A university library management model for students' learning support

Kulthida Tuamsuk [1], Kanyarat Kwiecien [1], Jutharat Sarawanawong [2]

[1] *Information & Communication Management Program, Khon Kaen University, Thailand*
[2] *Department of Library Science, Faculty of Humanities, Kasetsart University, Thailand*

Abstract

This research was aimed at developing a university library management model that would support students' learning. The research was conducted in three phases: 1) an investigation into the requirements of a university library service of instructors; 2) an investigation into the attitudes of librarians, libraries' administrators, and university's administrators toward the roles of library and policy concepts relating to university library management that promote students' learning; and 3) the development of a university library management model for students' learning support. A mixed research method was applied comprising qualitative and quantitative approaches. Data was collected using in-depth interviews and questionnaires. The findings have led to a proposal of a university library management model that supports student learning, which is comprised of five components: 1) management policy and system; 2) learning resources; 3) learning support services; 4) learning environments; and 5) the competency and roles of information professionals. It can be stated that studies into Thai university library implementation, strategic plans, and self-assessment reports under the quality assurance system showed that even though libraries have updated resources and services following changing situations in policies, technologies, and users' needs, there is no clear indicator that Thai university libraries have any strategy for acquiring roles to support students' learning with practical outcomes. Through reliable research work, this study into a model for university library management would result in a means of developing university libraries that truly supports university student learning based on information from instructors, library administrators, and librarians.

KEYWORDS

University library management, Library learning support, Thai university libraries

1. INTRODUCTION

Learning is a process derived from searching and acquiring new knowledge, from developing former knowledge that will result in human behavioral changes, practices that rely on knowledge and skills, and the expression of attitudes and values. Learning can be developed by means of education, learning, and reinforcement, and it is a life-long process for humankind (Birkenholz, 1999). Present university instruction emphasizes self-learning and researching, with the expectation that

1. Introduction

learners will be able to acquire knowledge on their own to meet individual needs, interests, and aptitudes. The duty of instructors is to promote learners' freedom in decision-making and applying their intellects to the fullest capacity, and to provide opportunities for learners to select their own learning activities. Instructors should train students to know how to learn independently from the beginning of university. It is the instructor's duty to introduce learning approaches, methods for searching learning sources, and how to cope with reporting work, so that learners can appropriately plan their learning. Learners can learn best when they want to learn. The intellectual capacity of each individual may not be the same; however, each person can learn if given time.

In learner-centered instruction – where learners acquire learning behaviors by themselves – the instructors must change their teaching behavior from sole knowledge-giving to active student participation. This can be achieved by posing questions that elicit thinking, thus motivating students to answer and learn. Questioning is an important means of instruction that leads to learners thinking. Besides using questions as an instructional approach, the introduction of information sources and explanation of their use is also essential. The library is where important self-learning resources exist; hence, a library should be managed in line with concepts such as administration, services, management of resources and learning media, the arrangement of atmosphere and facilities that enhance learning, and, not least, cooperation in the teaching and learning of the instructor. A perfect library, and a proper grasp of its role in enhancing learning and teaching, will contribute to the success of institutional student development.

The development of information technology, and the increasing use of resource materials to supplement, extend, and even replace lectures and seminars, has had a considerable impact upon libraries and the services they provide for users. Any approach to education adopted by an academic institution invariably affects the operations of the library (Arko-Cobbah, 2004). A literature review on the roles of university libraries that support students' self-learning demonstrated how librarians have direct roles in promoting students' independent learning. Independent learning places a greater strain on librarians as advisors and tutors (Brophy, 2005). Self-learning behavior increases students' need for information and enhances their researching habits; this means the information provided by the library may not be adequate. Thus, one duty of librarians is to consider broader uses of external information sources by means of cooperation with libraries in other universities, or by locating other existing and available information sources, and then managing the acquired sources effectively for the benefits of instruction (Goodall & Brophy, 1997). Additionally, information services in the library are also important. Librarians should improve existing services in order to best respond to students' self-learning. For example, the service time can be extended to facilitate learners, and various models of service can be incorporated that underscore both individuals' and group's needs. The library's atmosphere can be improved to support self-learning, the roles of librarians can be adjusted to support students' learning, and broad and easily accessible information resources and media inventory can be compiled to support curricular instruction.

Since the library is an essential part of student learning, it should be reformed so that it is able to integrate its roles into institutional educational reform. Users of the library should be the center of learning, just like in the so-called 'student-centered' approach. The library's management and services should be able to meet with the needs for information from various sources, and for information that can be analyzed and synthesized according to subjects' curricula. Since instructors have a major role in determining course content, activities, and instructional processes, cooperation between the librarian and instructor is vital for the development of the library as a source for learning support. However, studies into Thai university library management, strategic plans, and self-assessment reports under the quality assurance system of the Thai higher education system showed that even though libraries have updated resources and services following changing situations in policies, technologies, and users' needs, there is no clear indicator that Thai university libraries possess any strategy for acquiring roles to support students' learning with practical outcomes. Hence, this study, which was aimed at developing and proposing a university library management model that supports students' learning based on information from instructors, library administrators, and librarians in charge of services through reliable research approaches, would result in a means of developing university libraries that truly supports university student learning.

2. LITERATURE REVIEW

The main duty of university libraries is the support of instruction and research under each university's curricula. In general, university libraries are well prepared in terms of learning resources, tools and information technology equipment, and competent personnel to provide information services – all of which support learning. Learners at the university level are generally ready and keen to learn new things. As grown-up learners, university students are able to reason, understand, and explain abstract ideas effectively. The way learners at this level spend their time affects their learning and development: if the time is spent on things related to their field of study, academic outcomes will be improved. In general, learners are happy to meet and talk with instructors outside the class to ask questions they fail to understand, since it is leisure time that will not impinge on others' learning. They can also acquire knowledge in the library. The study by Liangjindathavorn (1997) showed that students' library use correlated with learning achievement. Students who study and research by themselves in the library and who are capable of appropriately using different services for their learning have been found to be able to manage their learning effectively, and usually show good academic records. Therefore, university libraries are important learning resources whose role in helping to intensively develop students' learning should be promoted. The library's working system should be adjusted in order to respond to the university's instructional requirement that accentuates learners as the center of learning, and to its policy of preparing graduates with learning skills and the capacity for life-long self-learning (Tuamsuk, 2004).

A university library is of interest to administrative academics, since it is a large and important organization with an impact on the university's achievements. In literature related to the history of library administration, there were several attempts to propose different models for library management. With the increasing size of libraries, the need for an official structure for control efficiency has become more obvious. Library hierarchical structure and scientific management are often found, in which two sections are generally structured: the technical section and the service section (and, in some cases, the administrative section) (Lynch, 1988). Sub-sectioning is based on the principle of operational clarity in each section. Personnel are placed to perform regular duties in each section. Although a link exists between each section according to the necessity for cooperation, cross-sectional operations are few, partly due to the number of orders and regulations within a library.Hummel (1987) noted that officially regulated administration not only has an effect on the conceptualization and implementation of people inside the organization, but also on the conceptualization and implementation of people outside the organization. Therefore, the library's image is one full of regulations and orders governing users.

Townley (1995) stated that although many university libraries accepted the reasons and necessities for alteration of their management model to meet external environmental changes – including users, institutional policies, and the use of technology in instruction – only a few libraries have changed their administrative structure. However, some university libraries have adopted a flat organizational structure, considering that the

2. Literature review

library provides information services, and is under a condition where technology should be used and users should be primarily considered. Thus, the organizational structure should not be divided, but each section should be linked together, acting as part of another. Library administration is emphasized on the library's goals and objectives more than internal power and order. University libraries have applied many modern organizational administration concepts, and many examples of university library management models have been proposed (Budd, 1998). One of these was the team-based model, which is popular among large-sized libraries with a complicated administrative structure and a number of library branches. A team-based administrative structure is applied to reduce operational steps and to enable an efficient response in a rapidly changing environment. The information institutional model conceptualizes a library as a component of the information environment of the whole related institution; that is to say, the library has to cooperate with other institutional units whose duty it is to produce and provide information services, e.g. university archives, educational technology centers, and computer centers. All such units are to respond to the institution's information policy. The third concept is an unbounded model, which still places importance on the traditional collection of information resources in the library, though integrated with other channels of information services for users. The education/research consultant model determines how librarians' roles are principally in teaching and researching. Librarians form a partnership with instructors to support the teaching and searching of research information.

Most Thai university library management is centralized in order to economize budget, manpower, and time. Administrative structure is hierarchical, and divided into different sections. A number of libraries have adopted modern administration to streamline operational efficiency; for example, total quality management (TQM), benchmarking, strategic management, or a balanced scorecard. In addition, modern library management has to facilitate common purposes and systematic thinking to enable personnel to initiate rather than follow suit; for instance, in planning or in setting targets together (Boonyakanchana, 1998). Based on the needs of universities to become a global leader or prototype of learning organizations, structuralizing the university library must be appropriate and in accordance with institutional implementation policies in order to enhance and support rapid institutional development in terms of quality, efficiency, and competitive competence (Poomvises, 2000).

Watson (2008) suggested that there are three areas of student support that benefit from taking an approach based on synthesis. The skills, attitudes, and behaviors of library staff are deeply affected by the technologies that we deploy and how we make them available, and also by the design and configuration of the environments in which they work. An approach based on synthesis which considers people, technology, and environments brings out synergies between these areas of investment. For example, a strategy of self-service can use the best technology available, but it is unlikely to succeed without staff that embrace it or physical space that enables easily understood access to, and use of, the facilities.

Furthermore, with the increasing quantity of information resources, the impact on library operations becomes obvious. The library is required to acquire more information resources to respond to users' needs without considering the resource quantity, since accuracy, completeness, and righteousness are more important. Therefore, modern libraries must be well prepared in terms of strategies to accommodate changing information resources. Implementation of modern libraries needs to be more proactive, and their services need to be correlated to the instruction of increasingly offered university programs. Acquisition and provision of information resources in the library has shifted from ownership, to assistance, to access to required information, and from the format of the information to the content users require. Hence, the library is placing more importance on electronic media sources, since they can be

2. Literature review

produced, disseminated, and retrieved rapidly without physical limitation. Users can benefit from the use of information within the library, and from any external information sources (Stueart & Moran, 2007).

Arko-Cobbah (2004) stated that the current shift of emphasis away from conventional teaching toward other learning strategies – especially student-centered learning – has increased the importance of libraries to their parent institutions. Librarians are expected to play various roles in furthering the aims of student-centered learning. For example, playing a role in the provision of information becomes more crucial, as it is bound to provide these resources to faculties and students at locations other than the library building. The library needs to carefully select and organize collections in all formats, network information resources, and then make them available in order to aid self-paced learning programs. The library is also expected to provide the necessary infrastructure to support instruction for all types of users, so as to enable them to meet their information needs. Another role is to foster partnerships and collaboration with personnel from other departments within the institutional set-up. Librarians must forge partnerships with teachers and faculty on all levels of education to bring about curricular restructuring and dynamic learning environments for students in the information age. The role of librarians will not only be limited to giving bibliographic instructions, but they will also be members of the team, along with faculty members and computer staff, who design networked course materials customized to serve individual student interests.

Since instructors take a prominent role in determining course content, activities, and teaching processes, cooperation between instructors and librarians in developing the library as a learning support source is a necessity. Besides this, library or institutional administrators take major roles in supporting the library to truly operate according to the needs of supporting students' learning at the institution. This research is therefore aimed at studying the roles of teachers, library and university administrators, and librarians, all of whom play a major part in piloting a good learners' supporting library model. The investigated research issues comprised library management policies and concepts in students' learning support, including learning resources and media, services and activities, physical environment that facilitates learning, and librarians and information professionals.

3. RESEARCH METHODOLOGY

A mixed method was applied, including qualitative and quantitative approaches. The research unit included libraries in four large public and autonomous universities in the northeast of Thailand, which had been purposively selected as representatives of university libraries. The research was conducted in three phases:

Phase 1: A study of the need for library services to support learning. The study was conducted quantitatively on the sample group taken from the four universities, which comprised 1136 instructors. A questionnaire was used as a survey tool to collect data. 914 questionnaire forms were returned (80.46%), and the data were analyzed into percentages, averages, and Chi-square.

Phase 2: A study of the knowledge and understanding of the library's roles, policy concept, and problems and obstacles in managing and providing university library services that promote students' learning. The study was divided into two parts: 1) the administrator groups – qualitative research was applied to university administrators responsible for library management and university library directors at the four universities. The data were collected by means of in-depth interviewing, and the answers were analyzed and synthesized by the descriptive method; 2) the librarian group – a quantitative approach was applied to 79 librarians at the four universities. A questionnaire was used as a survey research tool to collect the data. 75 questionnaires were returned (94.94%), and the data were computed into percentages, averages, and Chi-square.

Phase 3: The development and proposal of the university library management model to support students' learning. The results from Phase 1 and Phase 2 were developed into the model for university library management. The model was confirmed by means of quantitative and qualitative methods. The sample groups were selected from a number of library management specialists, educators in the field of library and information sciences, and librarians responsible for services in university libraries. The data were collected through interviews and survey forms. An index of consistency was derived from the data analysis.

4. CONCLUSION

4.1. Opinions of administrators, instructors, and librarians

4.1.1. Policy concept

In terms of the policy concept of supporting the library's role in developing students' learning processes, the administrators believed that the university should have a policy to enhance students' learning processes. Instructors should integrate learning processes into their instruction. The library should be developed until it is able to support such instruction by improving the infrastructure – including information technology, networking, e-learning, e-library, and library staff development – in order to be able to provide services that promote learning. The library organizational structure should be adjusted, with an emphasis on management efficiency. The organization should be flat, and management stages reduced. Team-based operations should be adopted in which achievement is underscored on the outcomes that reflect the university's targets. The library's roles should be shifted from technical-oriented to service-oriented ones.

4.1.2. Learning resources

In terms of learning resources, the administrators thought they should be in accordance with the instruction that users are the first priority. Librarians should work with instructors to provide learning media appropriate to the curricula. The library should be informed of all the programs offered in the university, and should analyze how far the existing learning resources respond to the institution and users. The library should take another role in developing and producing digital learning media. The learning resources should be compiled on the Internet by means of a web portal, which can be categorized into disciplinary areas of the university so that users can have easy access to online information. In general, the instructors and librarians want the library to acquire learning resources as required by the university's curricula and subjects (73.10% and 92.00% respectively); to provide databases that match the content of some subjects (72.30%, 85.30%); to provide learning resources from other sources that are in concordance with the curricula and subjects (71.70%, 82.70%); to produce knowledge bases that support the curricula and subjects, and compile these on the computer network of the library or of the university (68.50%, 81.40%); and to produce learning media that support self-learning (59.00%, 50.70%).

4.1.3. Learning support services

With regards to opinions on the learning support services, the administrators thought the library should take a more proactive role. This can be achieved by means of the customer relationship management strategy, in which services are provided for individuals or special groups. The library must acquire instructors' information and provide services accordingly, in the same way as business organizations serve their customers. Teaching and training on self-learning skills or information literacy skills should be given to students. Activities should be scheduled systematically and continuously to promote student learning activities in the library. Services should be provided to assist instructors in their teaching and research, with disciplinary classification according to the university's curricula and programs offered. The communication channels should be arranged so as to contact instructors and regularly inform them of news, and to facilitate maximal access to information. In general, the instructors and librarians want the library to involve in teaching and developing the students' information literacy skills (78.12% and 85.33%); to organize the students' learning promotion system and activities regularly and continually (71.33% and 93.33%); to build a communication system via several channels to provide information services to the users (68.05% and 69.33%); and to provide in-depth services, emphasizing individual and group users according to their profiles and interests (57.22% and 81.33%).

4.1.4. Physical environment

With reference to opinions on the physical environment that supports learning, the administrators believed that the surroundings and atmosphere of the university library can be designed based on the philosophy and concept of the university. In some cases, the university library may produce a more academic atmosphere through solemnity; peace and quiet within the knowledge and information resource surroundings is a target. At the same time, present learning and researching does not require one to visit the library, since modern information technology enables us to simply search and gain access to knowledge in the blink of an eye. Thus, many university libraries cannot avoid modernity and high-capacity information technology. For example, there are service zones incorporated into the library, such as the learning common zone and the information common zone, which attract instructors and students to use the library in a modern technological atmosphere that facilitates informal learning. Most of the instructors and librarians want the library to provide computers for searching information both in the library and outside (77.90% and 93.30% respectively); produce multiple and efficient information retrieval tools and equipment (75.00%, 93.30%); provide modern teaching media and equipment appropriate for self-learning and instruction in the library (65.80%, 70.70%); and provide a space or room at the library that facilitates the teaching of various styles (60.20%, 57.30%).

4.1.5. Capacity and role

In terms of opinions toward the capacity and role of information professionals, the administrators thought that besides knowledge in the field of library and information science, they should have knowledge on learner-centered instruction, as well as on programs offered in the university, in order to contact, assist, support, and cooperate with instructors to administer, give service, and manage information in each course. Therefore, university information professionals need a strong grasp of interpersonal skills, information technology, knowledge of marketing, learning sources, and information retrieval. Information professionals with a lot of experience may require continuing education in a certain field to provide specialized support to a particular group or field. In terms of librarians' roles, the administrators believe that librarians should be responsible for the entire process, not just specific areas. Their work should be a complete cycle from acquisition, organizing, services provided, and teaching students and instructors to search and use appropriate information. Most of the instructors and librarians believe that information professionals should take a role in implementing short-term training in information literacy or cooperating with instructors to teach information literacy topics in the course (61.00%, 77.40%); should introduce media or other learning sources relevant to course contents (56.90%, 82.70%); assist in or suggest information searches for producing instructional materials in different courses (56.50%, 66.60%); and advise instructors to prepare resources and learning media that exist in the library, which are in line with the course content (49.90%, 57.30) (Table 1).

4. Conclusion

Table 1. Opinions of instructors and librarians on the roles of university library for students' learning support.

	University library management model for student's learning support	Instructors N = 914	Librarians N = 75
1	**Learning resources**		
1.2	Preparation of the existing learning resources to support student's learning in accordance with the university curricula and courses instructions.	668 (73.10%)	69 (92.00%)
1.1	Acquisition of the external learning resources in accordance with the university curricula and courses instructions.	655 (71.70%)	62 (82.70%)
1.3	Development of the knowledge-based digital library for learning and teaching supports, especially the knowledge of university academic and research publications.	626 (68.50%)	61 (81.40%)
1.4	Development of the local databases in various subjects relating the courses offering in the university.	661 (72.30%)	64 (85.30%)
1.5	Development of the courseware/learning media for students' self-learning support.	540 (59.00%)	38 (50.70%)
1.6	Others (please specify)	6 (0.70%)	2 (2.67%)
2	**Learning support services**		
2.1	Involving in teaching and developing of self-learning skills or information literacy skills for students.	714 (78.12%)	64 (85.33%)
2.2	Organizing of the students' learning promotion system and activities regularly and continually, and correlate to contents or academically interesting issues and social changes.	652 (71.33%)	70 (93.33%)
2.3	Providing the in-depth services, emphasizing individual and group users according to their profiles and interests.	523 (57.22%)	61 (81.33%)
2.4	Building a communication system between the library and the users via several channels to provide regular information or warning, facilitating access and perception of information among faculty staff and students.	622 (68.05%)	52 (69.33%)
2.5	Providing the research support services to enhance instruction at graduate education levels and to develop research skills and learning at these levels, especially if the library is under the research-based university.	504 (55.14%)	56 (74.67%)
3	**Learning environments and infrastructure**		
3.1	Provision of space or room at the library that facilitates the learning and teaching of various styles.	550 (60.20%)	43 (57.30%)
3.2	Provision of modern teaching media and equipment appropriate for self-learning and instruction in the library.	601 (65.80%)	53 (70.70%)
3.3	Provision of high-capacity information technology to facilitate the use of and access to information both within and outside the library.	712 (77.90%)	70 (93.30%)
3.4	Provision of multi and efficient information retrieval tools and equipment to facilitate information seeking and acquiring of the users	685 (75.00%)	(70 (93.30%)
3.5	Others (please specify)	4 (0.40%)	2 (2.67%)
4	**Librarian and information professional roles**		
4.1	Collaborations with the faculty staff on course planning to incorporate the library roles in the course teaching and learning.	256 (28.00%)	33 (44.00%)
4.2	Advise the faculty staff and students on the course teaching and learning processes which relating to the uses of library for enhancing the students' learning outcomes.	456 (49.90%)	43 (57.30%)
4.3	Facilitating and suggesting learning resources and media that are in line with different programs and subjects of the university.	520 (56.90%)	62 (82.70%)
4.4	Assisting and recommending instructors and students in specific researching areas.	516 (56.50%)	50 (66.60%)
4.2	Implementing short-term training in information literacy or cooperating with instructors to teach information literacy topics in the courses.	558 (61.00%)	58 (77.40%)
4.6	Others (please specify)	5 (0.50%)	3 (4.00%)

4.2. Results of the model development

In developing and proposing a university library management model that supports student learning, the researchers conducted an analysis and synthesis of the results of Phase 1 and Phase 2, and drew a conclusion, before drafting the conceptual model of university library management according to the research conceptual framework. The quantitative data derived from analyzing the opinions of the instructors and librarians were considered when the scores were at a high level of above 70 per cent, or when they had an average raw score of over 3.50 (from the full score of 4). The qualitative data from the interviews with administrators were classified into categories, which, seen as they were opinions of most administrators, might match the quantitative data results or make an important set of additional data. The data from the two sets were classified appropriately in new groups based on academic viewpoints. Finally, the model of university library management for students' learning support was drafted. Following this stage, the drafted model was assessed by 15 specialists; then the results were analyzed by finding the index of consistency (IOC). The evaluation criterion stipulated by researchers was an IOC value of 0.8 and over. We thus obtained the *University Library Management Model for Students' Learning Support*, with five components (Fig. 1) as follows:

4.2.1. The management policy and system

1) Universities should have a teaching and learning policy that enhances students' learning, and which is the mechanism behind the building of graduates' qualities, so that they will be equipped with the skills for selflearning and life-long learning.

2) Universities should have a policy for the use of teaching techniques in different courses where students' learning is enhanced.

3) Universities should have a policy to enforce the use of the university library as the main learning source for university learners.

4) University libraries should have a target and strategy to provide services that clearly and concretely support students' learning, both for the short-term and in the long run.

4. Conclusion

5) University libraries' physical infrastructures should be developed alongside information technology that facilitates students' self-learning.

6) University libraries' management should be adjusted so that it becomes more flexible, with fewer operational steps, and which is able to respond quickly to changes to university policies and users' needs. For instance, the following approaches can be adopted, namely, team-based, service-oriented, cross-functional, or result-based operations that provide services that enhance learners' learning and support instructors' student-centered approach.

7) University libraries should have a structural mechanism and/or system in which the university community can vision the roles and importance of the library in supporting teaching and learning. This should clearly emphasize students' learning, and be as well accepted as other supporting units in the university, such as the teaching and learning support unit, the information technology support unit, or the student development unit.

8) University libraries should have a system and mechanism to promote and equip librarians and their teams with knowledge and competence, and which prepares them to enhance student learning.

4.2.2. Learning resources

1) The provision of learning resources must be in accordance with instructions, and be principally based on the consideration of users. Librarians must work with instructors to acquire learning resources in concordance with curricula. This may be done by surveying instructors' needs, or by considering the list of textbooks in the teaching plan of each subject.

2) Different formats of learning resources are provided, including printed media, electronics media, and digital media, with ease of

access to the contents that are in concordance with subjects, and available channels of access both inside and outside the library.

3) Knowledge bases are developed in the digital library that compiles learning resources supporting university curricular instruction, especially the learning resources that are academic work, research, or institutional repository.

4) Learning resources on the Internet are compiled systematically through a web portal that can be divided into different fields offered in the university, so that users can have greater access to worldwide information.

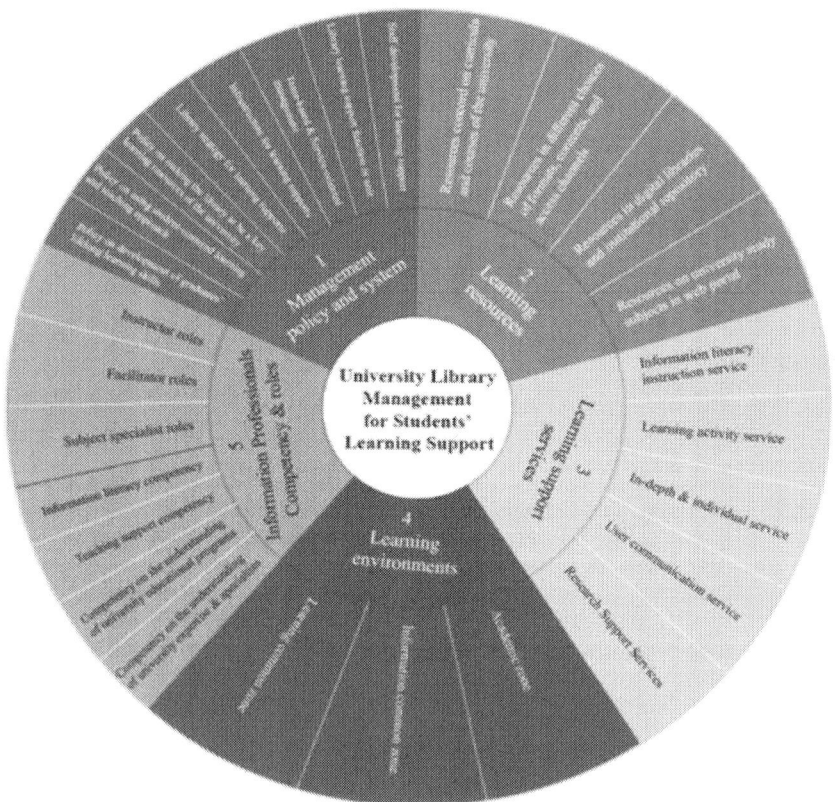

Figure 1. University library management model for students' learning support.

4.2.3. Learning support services

1) Teaching and developing self-learning skills or information literacy skills is provided for students systematically, with an emphasis on the development of selflearning processes.

2) The system and activities that promote students' learning are concrete, organized continually, and correlate to contents or academically interesting issues and social changes. The library should be dynamic and hold activities that continuously motivate students to learn.

3) In-depth services are provided, emphasizing individual and group users. Customer relation techniques can be applied to get to know users, and to build incentives, impressions, and confidence toward the services among users. Profiles of individual instructor users should be made in order to interrelate each instructor's interests, expertise, and subjects they are responsible for.

4) A communication system consisting of several channels should be built between the library and users to provide regular information or warning, and to facilitate access and perception of information among instructors. A question answering service or service for external requirements should be provided in order to extend the library to the university community.

5) Research support services should be provided to enhance instruction, and to develop research skills and learning at graduate levels, especially if the library is within a research-based university. For instance, advisory services can be offered on research methods and techniques, research analysis and synthesis methods, use of programs to write references, statistical data analysis, research report writing, and academic article writing.

4.2.4. Learning environments

The learning environment should consist of three components:

1) Learning common zone: The objectives of the learning common zone are to provide a learning area with an atmosphere that attracts instructors and students to use the library for information, entertainment, socialization, and joining learning activities. The area can be designed as a newspaper/magazine/comic corner, a coffee corner, a garden corner, an exhibition corner, a public relation corner, or 'edutainment'.

2) Information common zone: This zone is provided as a service area equipped with modern information technology that enables users to have access to multiple knowledge sources. The use of information technology itself gives freedom to users in their learning acquisition. The major technology that composes this service is a high potential one that enables users to use the Internet via a wireless network with modern computer software and hardware. Users will have access to both internal and external library information sources. Services for the retrieval of materials, answering of questions, and teaching or supervising the use of technological devices and equipment are incorporated around the clock.

3) Academic zone: The objective of this zone is to offer an area for searching for academic and research information, with an emphasis on the ampleness of such information resources' concordance with university curricula and programs. Resources with an academic atmosphere are provided to facilitate learning. The major components in the zone may include question answering service, searching assistance, research support, reference compilation, a space or corner to situate individual and group researching tools, or a space or room for different group sizes for lecturing or discussion.

4.2.5. Information professionals

The information professional component is divided into two parts: the required competencies and operational roles. The required competencies of information professionals working in university libraries include:

1) Information literacy competency e with an aim to enable students to efficiently develop their own learning skills by means of multiple learning sources;

2) Teaching support competency e with an aim to instill knowledge and understanding of instructional processes that promote learning among students, especially on the part of information professionals to support teachers in this respect;

3) Competency in the understanding of university educational programs, including the aim of learning process development stated in the curriculum of each subject;

4) Competency in the understanding of university expertise and specialties e requiring support from the library in the form of in-depth services, especially in the research areas continuously conducted, or in programs offered at graduate levels.

5) The operational roles to be taken by information professionals to promote students' learning include:

6) Instructor roles e information professionals should be competent in implementing short-term training in the library by cooperating with instructors to teach certain topics, or offer some courses with related fields or university units. Then the curriculum can be produced online or on the university's e-learning component;

7) Facilitator roles e information professionals should facilitate learning that is in line with the different programs and subjects of the university. The operation should be systematic, with collaboration

from instructors, compilation that eases access, introduction for usage through different channels, and continuous updates;

8) Subject specialist roles e information professionals should assist and recommend instructors and students in specific researching areas. In this regard, they should receive training in order to take on entire responsibilities. They must also possess an attitude that all functions mean services: the library does not classify information professionals to be working only on one area as they formerly were, e.g. procurement, sectioning analysis, resource inventory, services, etc. The content of services should be underscored.

5. DISCUSSION

This research study proposes a university library management model that supports students' learning. The model is comprised of five major components, namely, policy and management, learning resources, learning support services, physical environment supporting learning, and information professionals. It has been shown that policy and management is the primary component of the university library management model, since library operations are principally under a university's policies. If university and library administrators see the vitality of the library as students' learning support mechanism – and hence stipulate this as a major policy – instruction and library operations will follow suit. Instructors will run their courses in such a way that promotes self-learning. The library itself will have to provide efficient information resources, systems, and information resource-retrieving tools and services that are in line with users' needs in order to facilitate maximal access to information. The belief that policy and management systems governing university libraries is vital correlates to Punpruk's (2004) work on student-centered instructional approaches under institutional indicators for the development of thinking and knowledge-building capacities. The study revealed that a factor involving administrators leads to instructional success in which students are centered, as stated in the institutional indicators. The success of student-centered instruction according to institutional indicators necessitates the vision and leadership of administrators, especially those with academic responsibilities. The success also depends on a participatory style of administration and student-centered instruction. This notion correlates to the study of Pakawatchai (2004), who found that determined, devoted administrators, who are aware of learner-centered teaching and who are keen to attain academic leadership, often drive the institution to successful reform. The library necessitates managerial alterations so that it becomes more effective and streamlined, with unnecessary steps reduced and hence more rapid services. The system in which service expertise is adopted is incorporated whereby an information professional can cope with all functions in the library, and is able to take responsibility in a specific field. In this way, information professionals will be trained to acquire expertise in a certain field, and are therefore able to provide maximal services to users. For instance, some university libraries in the United States have shifted to a flat organization structure after seeing that the library is a service organization in which the use of technology is a must and users are a priority. Thus, organizational management should be one that is not sectioned, but should have clear independence, and interrelations of components that can be a part of the others. Library management gives greater importance to the objectives and targets of the library than to power and control over organizational order (Budd, 2005; Townley, 1995).

With respect to physical environment that enhances learning, the findings show that the library formerly emphasized the physical environment only for academic purposes. However, when the needs of users and human learning behaviors change, the library needs to adjust the internal environment in accordance with external situations. The results of the study are in line with the study by Juceviciene and

Tautkeviciene (2004), who investigated the learning environment of the library as an educational environment of the university. Their study indicated that students at all levels use the library to supplement their education. Students utilize information and service resources both in the forms of printed matters and digital information, as well as face-to-face and virtual services. Besides this, the present users' needs are not limited to learning support only. If life-long learning is a policy, then a venue for serving informal learning and the use of information resources should be provided so that the atmosphere will be more relaxed, not tense as it used to be, in order to attract more students. Fox and Stuart (2009) and Sinclair (2007) also proposed the concept called Common 2.0, which consists of five components: 1) an open system – the library is quiet and peaceful, suitable for individual work and encouraging of learning, but at the same time allowing users to exchange opinions; 2) a free system – students are able to bring their laptops to the library and use a free network in order to retrieve information inside and outside the library; 3) a comfortable place – the library arranges learning zones according to learners' categories and learning styles. A place for common learning is provided for teachers, students, and people in general. The design and arrangements of desks, chairs, and furniture – as well as the lighting system – offers comfort and convenience for use; 4) an inspiring place – the positioning and design of tools and equipment must be appropriate for use, yet also beautiful, elaborate, and creative in order to inspire students and other users to learn; 5) practicality – the library should provide areas for students to learn, as explained in various learning theories, e.g. constructivist theory. Thus, the library should support circular and informal learning through the provision of services, supplementary equipment for students' work including Internet connection accessories, advice for library usage, information technology and communication for knowledge research, and information technology assistance.

The capacities of information professionals have to be strengthened. Besides their knowledge in library and information sciences, they have to be trained in the teaching of information literacy to students. Information professionals need to know and understand the academic directions and specific expertise of their university in order to locate the true needs of users; they can then prepare and provide information resources accordingly. This correlates to the study by Tanloet and Tuamsuk (2011) on the core competencies of information professionals in the next decade (2010–2019). Information professionals in the next decade need to be capable in terms of basic knowledge in information sources, knowledge and information management, information services, organizational management, research and user study, and continuing education and life-long learning. A study by Khoo (n.d.) also looks into this; skills in teaching, training, and giving advice are necessary for the new generation of information professionals. Therefore, information professionals working in a university library where students' learning is emphasized in order to respond to a national education reform policy need to be educated in any field they are lacking in, so that they can effectively support student learning.

6. RECOMMENDATIONS

The research findings lead to the recommendation that university library management in Thailand should aim to change their roles in accordance with the university's vision so as to enhance student-centered learning and life-long learning strategies.

6.1. Policy for the university

The research findings indicate that the university's setting of graduates' abilities in terms of the skill of life-long learning should be linked with the library's mission. The university should have a mechanism and system to transfer the policy to the library, and the library in turn should exhibit a clear role in learning support with broadly recognized operational plans among curricular administrators, instructors, and students. The work of the library in this respect has to be monitored and evaluated intensively and concretely, and considered as part of the library's achievement indicators.

6.2. Implementation recommendations

The research findings reflect major components in management and operations that will compel the library to support the learning of learners. There are many aspects from which the library can select strategies for the improvement of work that encourages learning, which should also be appropriate to the strengths and weaknesses of each library. Programs that should be implemented include: 1) the development of information professionals so that they have knowledge and understanding of the learner-centered approach. Specifically, they should understand which learning activities relate to the library's roles, be able to read teaching plans, and analyze how much an instructor's expectation for learning involves the library; 2) the learning resources in the library should be improved and developed into a new style, new collections should be provided or new webs formed, so that instructors and students can see how much and how far the library is prepared to support learning; 3) service packages that enhance learning are continuously advertised. Libraries must see the importance of conscious building, perception, and acceptance of new things among the university's people, and not only support the previous concept which saw learning as a routine operation.

REFERENCES

1. Arko-Cobbah, A. (2004). The role of libraries in student-centered learning: the case of students from the disadvantaged communities. International Information and Library Review, 36(3), 263–271.
2. Birkenholz, R. J. (1999). Effective adult learning. Danville, IL: Interstate Publishers.
3. Boonyakanchana, C. (1998). New thinking for the future of library. Information, 5, 14–18.
4. Brophy, P. (2005). The academic library (2nd ed.). London: Facet Publishing.
5. Budd, J. M. (2005). The changing academic library: operations, cultures, environments. Chicago: ACRL, Publications in Librarianship.
6. Budd, J. M. (1998). The academic library: Its context, its purpose, and its operation. Englewood, CO: Libraries Unlimited.
7. Fox, R. & Stuart, C. (2009). Creating learning spaces through collaboration: how one library refined its approach. Educause Quarterly, 32(1). Retrieved from http://www.educause.edu/EDUCAUSE+Quarterly/EQVolume322009/EDUCAUSEQuarterlyMagazineVolum/163844.
8. Goodall, D. & Brophy, P. (1997). A comparable experience? Library support for franchised courses in higher education (British Library Research and Innovation report: 33). Preston: University of Central Lancashire.
9. Hummel, R. P. (1987). The bureaucratic experience (3rd ed.). New York: St. Martin's Press.
10. Juceviciene, P. & Tautkeviciene, G. (2004). The library learning environment as a part of university educational environment. In Paper presented at the European conference on educational research, 22–25 September, University of Crete. Retrieved from http://www.leeds.ac.uk/educol/documents/00003737.htm.

11. Khoo, C. S. (n.d.). Competencies for new era librarians and information professionals. Retrieved fromhttp:// www.lib.usm.my/elmu-equip/conference/ Documents/ ICOL%202005% 20Paper% 202%20Christopher%20Khoo.pdf.
12. Liangjindathavorn, O. (1997). A study of relationship between library use behaviors and study performances of UbonRatchathani University students. UbonRatchathani: The University Library.
13. Lynch, B. P. (1988). Changes in library organization. In A. Woodsworth, & B. Wahlde (Eds.), Leadership for research libraries (pp. 67–78). Scarecrow Press: Metuchen, NJ.
14. Pakawatchai, S. (2004). A research report on resource-based learning for the development of students' living knowledge and skill. Bangkok: Faculty of Education, Chulalongkorn University.
15. Poomvises, P. (2000). Model of academic library in the future. In Paper presented at the 18th seminar of Thai academic library cooperation on academic library in the millennium. 26–27 October, Bangkok.
16. Punpruk, S. (2004). A research report on student-centered learning and teaching approach for students' skill development in knowledge thinking and construction based on the performance indicators of educational institution. Khon Kaen: Faculty of Education, KhonKaen University.
17. Sinclair, B. (2007). Commons 2.0: library spaces designed for collaborative learning. Educause Quarterly, 30(4). Retrieved from http://www.educause.edu/EDUCAUSE+Quarterly/EDUCAUSEQuarterlyMagazineVolum/Common20Library SpacesDesigned/162265.
18. Stueart, R. D. & Moran, B. B. (2007). Library and information center management (7th ed.). Westport, CT: Libraries Unlimited.
19. Tanloet, P. & Tuamsuk, K. (2011). Core competencies for information professionals of Thai academic libraries in the next decade (A.D. 2010–2019). International Information & Library Review, 43(3), 122–129.

20. Townley, C. T. (1995). Designing effective library organization. In G. B. McCabe, & R. J. Person (Eds.), Academic libraries: Their rationale and role in American higher education. Westport, CT: Greenwood Press.
21. Tuamsuk, K. (2004). Resource-based learning and teaching approach. Journal of Innovative Learning & Teaching, 1(2), 1–6.
22. Watson, L. (2008). It's not about us: it's about them. In M. Weaver (Ed.), Transformative learning support models in higher education (pp. 1–18). Facet Publishing: London.

INDEX

A

Academic libraries, 27, 48, 60, 146, 148, 283, 307, 341

B

Bachelor's Degree, 228

Bulgaria, 227, 228, 230, 231, 241, 249

C

Campus-Wide Initiatives, 285

collaboration, 12, 14, 21, 26, 63, 176, 180, 182, 197, 216, 217, 219, 229, 250, 283, 287, 288, 289, 292, 296, 297, 298, 300, 303, 306, 307, 308, 309, 321, 333, 339

Collaboration, 176

Collections assessment, 174

D

Digital era, 146

Disciplinary knowledge, 146, 150

E

e-books, 113, 114, 115, 116, 117, 118, 119, 120, 121, 122, 124, 126, 128, 129, 130, 131, 133, 134, 135, 136, 137, 140, 141, 142, 143

Education, 60, 79, 84, 90, 91, 109, 170, 185, 192, 193, 199, 200, 208, 227, 228, 229, 230, 233, 240, 251, 252, 261, 286, 287, 308, 340

F

fourth paradigm, 2, 26

G

Generic skills, 146, 147, 152

J

Job advertisements, 79, 146, 170

Job hunting, 60

L

learning commons, 283, 284, 285, 286, 287, 288, 290, 291, 292, 293, 294, 295, 296, 297, 298, 299, 300, 301, 303, 304, 305, 307, 308, 309, 310, 311

Library instruction, 92, 105, 174, 182, 282

Longevity, 180

P

partnerships, 85, 101, 111, 177, 234, 250, 283, 284, 287, 288, 289, 290, 292, 295, 303, 321

Personal competencies, 146, 153

Q

Quality Control, 46, 253

R

RESEARCH DATA MANAGEMENT, 1

Romanian Academic Libraries, 30

S

Stereotypes, 29, 30, 32, 33, 35

Strategic Management Model, 45, 46

SULSIT, 227, 228, 230, 231, 232, 234, 237, 241, 242, 243, 244, 246, 250

T

Thai university libraries, 314, 317

U

University library management, 314, 330

V

Value of library services, 174

W

wider study, 145, 147, 157, 162, 164, 166, 167, 168